Impact of Artificial Intelligence on Society

The book presents a comprehensive and interdisciplinary exploration of the impact of AI on various sectors of society to foster a greater understanding of the opportunities and challenges presented by this transformative technology. It explores the impact AI has had on varied sectors of society, including healthcare, education, the workplace, and the economy. It provides a holistic view of this fast-growing technology by critical study of the possible benefits and drawbacks linked with the application of AI in many industries. The book also examines the ethical, social, and economic implications of AI and the potential risks and challenges associated with its use.

- Focuses on the future influence of AI, providing insights into how it could disrupt several industries and change the way we live, work, and connect with one another.
- Explores how AI can be used to tackle global issues such as climate change, food security, and public health concerns.
- Offers case studies and specific examples of how artificial intelligence is being employed in many industries, covering both successes and failures.
- Investigates cutting-edge technology breakthroughs in AI and how they can be used to improve efficiency, productivity, and performance across multiple industries.
- Understands the limitations and potential biases of artificial intelligence, as well as the significance of human monitoring and accountability.

The book is intended for researchers, practitioners, policymakers, and students who are interested in understanding the nature and role of AI with regard to different sectors of society.

Impact of Artificial Intelligence on Society

Edited by

Sumit Tripathi
Joanna Rosak-Szyrocka

CRC Press is an imprint of the
Taylor & Francis Group, an **informa** business
A CHAPMAN & HALL BOOK

Designed cover image: PopTika/Shutterstock

First edition published 2025
by CRC Press
2385 NW Executive Center Drive, Suite 320, Boca Raton FL 33431

and by CRC Press
4 Park Square, Milton Park, Abingdon, Oxon, OX14 4RN

CRC Press is an imprint of Taylor & Francis Group, LLC

ISBN: 978-1-032-64324-3 (hbk)
ISBN: 978-1-032-64449-3 (pbk)
ISBN: 978-1-032-64450-9 (ebk)

DOI: 10.1201/9781032644509

Typeset in Times
by SPi Technologies India Pvt Ltd (Straive)

Contents

Preface

The assimilation of Artificial Intelligence (AI) into several domains of contemporary society signifies a turning point in the history of technological progress. The goal of this book, *Impact of Artificial Intelligence on Society*, is to help readers understand how AI is affecting various facets of human life. It is critical to analyse and fully appreciate the complex implications AI has on our shared future, given that we find ourselves at the nexus of previously unimaginable technology capabilities and societal ramifications. This book began with the realization that AI has significantly more transformational capacity outside of its technological realm. Artificial intelligence has implications for social, ethical, political, and economic spheres as it permeates industries, economies, and everyday life. This collection of observations, analyses, and empirical research aims to clarify the subtleties of AI's influence by avoiding exaggeration and conjecture and instead basing conversations on factual findings. With a critical and progressive lens, this book seeks to further a comprehensive knowledge of the potential and problems associated with integrating AI into our global social fabric.

We have brought together responses from specialists in the domains of technology, economics, sociology, and ethics to promote a holistic analysis of the implications of artificial intelligence. Through the synthesis of multiple viewpoints, we strive to create a comprehensive story that appeals to readers of different backgrounds while maintaining the in depth analysis. It is our aim that this book will be a useful tool for people, governments, and academics as AI develops and spreads into new areas, encouraging a critical and informed discussion about the enormous social ramifications of AI. To sum up, *Impact of AI on Society* hopes to serve as a scholarly compass that leads readers through the complex terrain that AI has carved out, encouraging a thorough comprehension of its implications on the dynamics of our globalized society.

About the Authors

Dr. Sumit Tripathi is an accomplished associate professor at the Goa Institute of Management in Goa, India, bringing over 15 years of academic and research experience to his role. He earned his post-doctorate from Linkoping University, Sweden, and completed his PhD at the Indian Institute of Technology (BHU), Varanasi. Dr. Tripathi's research interests encompass Artificial Intelligence, Machine Learning, Deep Learning, Computer Vision, and Biomedical Image Processing. His expertise is evident through numerous high-impact research papers published in esteemed conferences and journals, as well as patents in Biomedical Image Processing using Deep Learning-based Techniques. As a lifetime member of the Indian Society of Technical Education (ISTE), Dr. Tripathi is dedicated to advancing the fields of engineering education and research.

Dr. Joanna Rosak-Szyrocka is assistant professor, Erasmus+ coordinator at the Faculty of Management, Czestochowa University of Technology, Poland. She specialized in the fields of digitalization, industry 5.0, quality 4.0, education, IoT, AI, and quality management. She completed a research internship at the University of Zilina, Slovakia, and at Silesia University of Technology, Poland. She has been a participant in multiple Erasmus+ teacher mobility programs: Italy, UK, Slovenia, Hungary, Czech Republic, Slovakia, and France. She cooperates with many universities both in the country (University of Szczecin, Rzeszów University of Technology, Silesian University of Technology) and abroad (including the University of Tabuk, Saudi Arabia; Széchenyi István University, Hungary; University Faisalabad, Pakistan; University of Humanities, China, University of Technology Sydney, Australia; Bucharest University of Economic Studies, Romania; and Federal University Dutse, Nigeria). She has also served as a member of the editorial boards and reviewer boards for several prestigious journals.

List of Contributors

Irfan Ahmed
University of Engineering & Technology Peshawar
Peshawar, Pakistan

Shubhra Bhatia
Bhopal School of Social Sciences
Bhopal, India

Cristina Dumitru
The National University of Science and Technology POLITEHNICA Bucharest, Pitești University Centre
Romania

S. Gayathri
Rajalakshmi School of Business
Chennai, India

Arif Hussain
University of Engineering & Technology Peshawar
Peshawar, Pakistan

Abid Iqbal
University of Engineering & Technology Peshawar
Peshawar, Pakistan

Muhammad Abeer Irfan
University of Engineering & Technology Peshawar
Peshawar, Pakistan

Almula Umay Karamanlıoğlu
Başkent University
Turkey

Amaad Khalil
University of Engineering & Technology Peshawar
Peshawar, Pakistan

Bhupendra Kumar
School of Computer Science and Applications, IIMT University
Meerut, UP, India

Nawnit Kumar
Sri Guru Tegh Bahadur Institute of Management & IT
New Delhi, India

Satish Kumar
Department of Information Technology
BGSB University Rajouri
J&K, India

Shalabh K. Mishra
Bharati Vidyapeeth's College of Engineering
New Delhi, India

Anand Nayyar
School of Computer Science
Duy Tan University
Da Nang, Viet Nam

Ramesh Nuthakki
Atria Institute of Technology
Bengaluru, India

Durdana Ovais
The BSSS Institute of Advanced Studies
Bhopal, India

Peterson K. Ozili
Central Bank of Nigeria
Nigeria

Pallavi Patwari
School of Liberal Arts and Management Studies
P P Savani University
Gujarat, India

Harshini Reddy Penthala
BVRIT HYDERABAD College of Engineering for Women
India

Joanna Rosak-Szyrocka
Czestochowa University of Technology Faculty of Management Department of Production Engineering and Safety
Poland

V. Selvalakshmi
Department of Management Studies
Velammal College of Engineering and Technology
Madurai, India

Kewal Krishan Sharma
School of Computer Science and Applications
IIMT University
Meerut, UP, India

Vikas Sharma
School of Computer Science and Applications
IIMT University
Meerut, UP, India

M. Shanmuga Sundari
BVRIT HYDERABAD College of Engineering for Women
India

Sumit Tripathi
Big Data Analytics
Goa Institute of Management
Goa, India

Aparna Vajpayee
School of Liberal Arts and Management Studies
P P Savani University
Gujarat, India

Tarun Kumar Vashishth
School of Computer Science and Applications
IIMT University
Meerut, UP, India

Seema Yadav
Department of Education Bhopal School of Social Sciences
Bhopal, India

Introduction

The *Impact of Artificial Intelligence on Society* presents a comprehensive exploration of AI's transformative role across diverse sectors. Chapter 1 investigates AI's disruptive potential in medical image segmentation, comparing traditional methods with AI-based approaches to enhance diagnostic accuracy and patient care. In Chapter 2, the applications of transformer models in medical imaging are reviewed, showcasing their ability to revolutionize tasks such as image classification and object detection. Chapter 3 explores the future of learning in higher education, highlighting adaptive learning systems based on language generative models to personalize learning experiences and optimize curriculum design. Moving to AI integration in higher education, Chapter 4 navigates through opportunities and ethical considerations, emphasizing responsible AI use for improved educational outcomes. Chapter 5 discusses how AI can revolutionize higher education by improving learning outcomes and operational efficiency in universities. Chapter 6 delves into AI's role in academic research, discussing advances and challenges in data analysis, predictive modeling, and interdisciplinary collaboration. Continuing the discussion on AI in education, Chapter 7 explores AI-enhanced content creation and personalized learning experiences to foster greater student engagement and comprehension. Chapter 8 investigates the interplay of educational social media usage, procrastination, and well-being in digital learning environments. Chapter 9 examines advancements and challenges in fraudulent message detection, emphasizing AI's role in mitigating online threats. Focusing on healthcare, Chapter 10 explores IoT-based telehealth systems for remote patient monitoring and diagnosis. In Chapter 11, AI's potential for social good is highlighted, addressing disaster relief, poverty alleviation, and environmental sustainability through innovative AI-driven solutions. Lastly, Chapter 12 explores digital innovations for increasing financial inclusion, showcasing AI's role in empowering underserved populations and fostering economic growth. Through these chapters, readers gain insights into AI's multifaceted applications, its potential to drive positive societal impact, and the challenges and opportunities inherent in its adoption across industries.

1 Disruptive Innovation in Medical Image Segmentation

A Comparative Study of Traditional and AI-Based Approaches

Sumit Tripathi
Goa Institute of Management Poriem, Sanquelim, India

Joanna Rosak-Szyrocka
Czestochowa University of Technology, Poland

1.1 INTRODUCTION TO SEGMENTATION

In the expansive domain of present-day healthcare, the fusion of technology and medicine has yielded unparalleled opportunities for transforming the processes of diagnosis, treatment, and patient care. The subject of medical image segmentation has witnessed significant improvements in recent years, which have had a dramatic impact on the field of synergy. This chapter explores the complex realm of pixel-by-pixel analysis, focusing on the significant role that medical image segmentation plays in the advancement of precision medicine (Balafar et al. 2010). The task of medical image segmentation involves the division or partitioning of medical images into distinct sections that hold significance for diagnosis and interpretation (Abler, Rockne, and Büchler 2019). This technique enables the segregation and analysis of distinct structures or abnormalities present in the images, regardless of their origin from magnetic resonance imaging (MRI), computed tomography (CT), ultrasound, or other imaging modalities. This procedure forms the basis for numerous clinical applications, ranging from the detection of tumors in radiological scans to the segmentation of anatomical structures for the purpose of surgical planning. The significance of medical image segmentation is mostly attributed to its ability to enhance individualized healthcare through the utilization of data-driven approaches. This technology enables healthcare practitioners to make well-informed judgments, develop customized treatment strategies, and accurately forecast patient outcomes

DOI: 10.1201/9781032644509-1

with exceptional accuracy (Bauer et al. 2013). The significance of this technique on patient care cannot be understated, as it is utilized for many purposes such as outlining malignant masses, distinguishing neuronal structures in brain scans, and segmenting cardiac regions for accurate diagnosis. The objective of segmentation is to demarcate the borders of objects by assigning a distinct label to each pixel or voxel. Pixels or voxels that possess identical labels exhibit comparable attributes or belong to the identical classification (Tripathi and Sharma 2021).

Image segmentation can be categorized into three primary methods: manual segmentation, semi-automatic segmentation, and fully automatic segmentation. Manual segmentation involves radiologists and doctors with expertise who delineate specific regions of interest (ROI). However, this approach is time-consuming and impractical for large datasets, and it can introduce significant variability. Semi-automatic methods combine manually designed elements with human operator intervention during processing stages. The emergence of fully automatic techniques, driven by the growing importance of computer-based systems, allows for complete control of feature extraction and segmentation. Deep learning, in particular, has played a crucial role in developing entirely automated methods capable of producing reliable and precise results. In the field of biomedical image segmentation, deep learning-based methods are attracting significant interest, particularly for diagnosing pathological tissues (Tripathi et al. 2021). Addressing the challenge of limited medical data availability has become a priority in the design of these networks. These approaches can autonomously extract features and learn. Deep learning-based architectures for image processing primarily employ supervised learning approaches.

This chapter will undertake a thorough examination of the fundamental principles, methodology, and practical applications pertaining to the process of medical image segmentation. The present chapter explores the progression of algorithms and approaches that have undergone development over time, enabling the interpretation and extraction of useful information from images with unparalleled precision.

1.2 CLASSICAL METHODS OF MEDICAL IMAGE SEGMENTATION

The segmentation of medical images is an essential task within the field of medical image analysis, as it holds significant importance in various areas like disease diagnosis, treatment planning, and medical research. Traditional approaches to medical segmentation of images often depend on existing methodologies in the fields of image processing and computer vision. The primary purpose of segmentation is to partition the objects inside an image. In the context of medical image segmentation, the objective is twofold: to examine the anatomical structure and to identify the Region of Interest (ROI), which involves locating tumors, lesions, and other irregularities. Tissue volume measurement is employed to assess the growth of tumors, as well as the reduction in tumor size resulting from treatment. Additionally, it aids in treatment planning before the administration of radiation therapy, namely in the computation of radiation dosage. The categorization of methods for segmenting the image can be broadly outlined as follows:

1.2.1 Methods That Rely on Gray-Level Features

These are the methods that take in to account the gray-level features for segmentation of medical images. Histogram-based segmentation is one such approach.

1.2.1.1 Histogram-Based Method for Segmentation

Within the field of image processing, a histogram serves as a visual representation that illustrates the distribution of pixel intensity values within an image. The provided data pertains to the relative brightness or darkness of pixels within different regions of the image. Through the examination of the histogram, one can acquire valuable knowledge pertaining to the contrast, brightness, and distribution of objects within the image. Histogram-based segmentation involves the analysis of pixel intensity values in order to identify and delineate distinct regions or segments (Caraiman and Manta 2014). One commonly employed technique is thresholding, which involves the selection of a threshold value to categorize pixels as belonging to a particular segment based on whether their intensity values exceed or fall below this threshold. In addition, adaptive thresholding can be employed, wherein the determination of the threshold value is contingent upon the localized attributes of the image. The thresholding function can be expressed by Equation 1.1:

$$u_{x,y} = \begin{cases} 1 & w_{x,y} \geq \theta \\ 0 & w_{x,y} < \theta \end{cases} \tag{1.1}$$

The output pixel at coordinate (x, y) is denoted as $u_{x,y}$. It is determined by the pixel value of the input image, $w_{x,y}$ and the threshold value, θ. This thresholding operation holds good for binary segmentation. Figure 1.1 shows the results after thresholding operation is performed over the brain and skin cancer images.

1.2.2 Methods That Rely on Edge Detection for Segmentation

Edge detection algorithms are a prevalent approach to image segmentation. These methods concentrate on detecting image borders or edges, which can then be utilized to segregate objects or regions of interest (Xu et al. 2021). The edge detection approach identifies differences in properties such as gray levels or colors, which frequently represent transitions between various regions. This method effectively divides an image based on object boundaries by segmenting it by pinpointing these transition zones. Edge detection-based methods are implemented in the following steps:

a) **Edge Detection**: To locate edges in an image, use a derivative operator.
b) **Edge Strength Measurement**: Determine the strength of detected edges by measuring the amplitude of gradient.
c) **Edge Pruning**: Remove weak edges by keeping only those whose magnitude exceeds a predefined threshold (T).

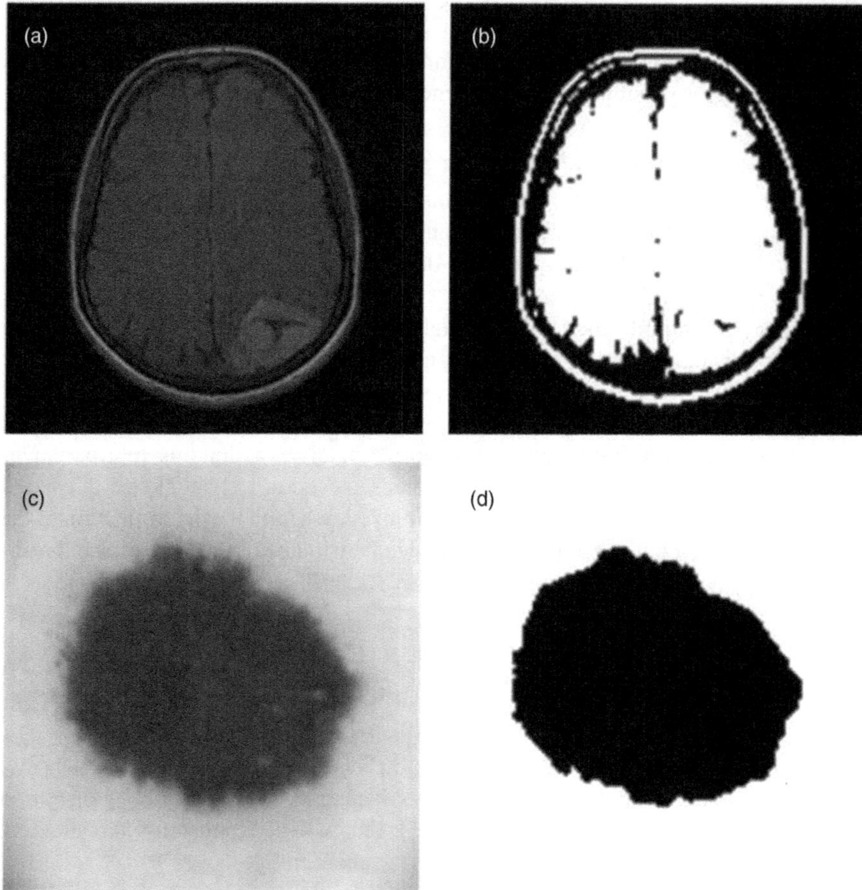

FIGURE 1.1 Results of histogram-based thresholding: (a) and (c) are original brain tumor and skin cancer images, (b) and (d) are segmented results of brain and skin cancer images.

d) **Crack Edge Analysis**: Determine the positions of crack edges and decide whether to keep or reject them based on the confidence of their neighboring edges.
e) **Closed Boundary Identification**: Repeat steps 3 and 4 with different threshold settings to identify closed borders and segment the image.

Prewitt, Sobel, Roberts (1st derivative type), Laplacian (2nd derivative type), Canny, and Marr-Hildreth, Edge Tracing edge detectors are among the gradient (derivative) function-based edge detection operators available. The goal of edge-based segmentation methods is to create object boundaries by connecting identified edges to form an edge chain. A thresholding operation is used to remove false or weak edges during this process. Edge-based segmentation results are presented in Figure 1.2 over brain and skin cancer images, respectively.

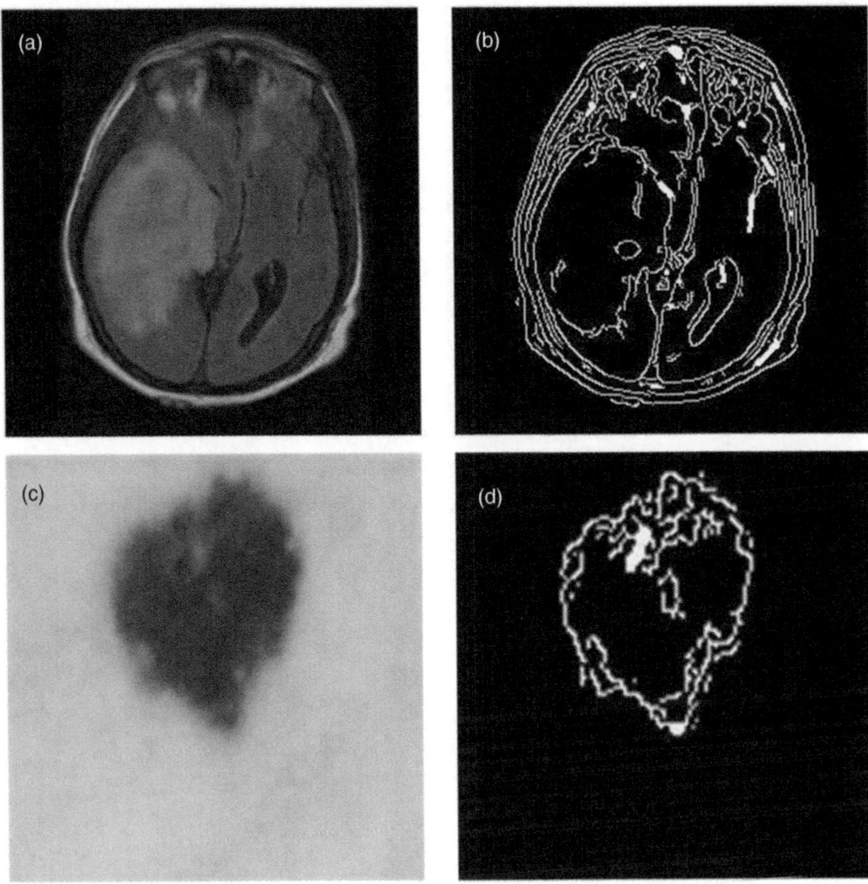

FIGURE 1.2 Results of edge-based segmentation (a) and (c) are original brain tumor and skin cancer images, (b) and (d) are segmented results of brain and skin cancer images.

1.2.3 Methods That Rely on Region Characteristics for Segmentation

A method for segmenting an image into regions or segments based on specific attributes or features is called region-based segmentation, and it is used in computer vision and image processing (Liu et al. 2021). Gathering together groups of pixels or regions in an image that have similar characteristics—like color, texture, intensity, or other visual aspects—is the aim of region-based segmentation. Numerous applications, such as object recognition, image analysis, and computer vision tasks, depend on this process. An image is a composite made up of several different regions, each acting as a separate component. When these separate parts or areas are put together, they make up the entire image. Equation 1.2 can be used to represent the relationship between regions:

$$X_a U\, X_b U\, X_c U \ldots\ldots UX_z = I \tag{1.2}$$

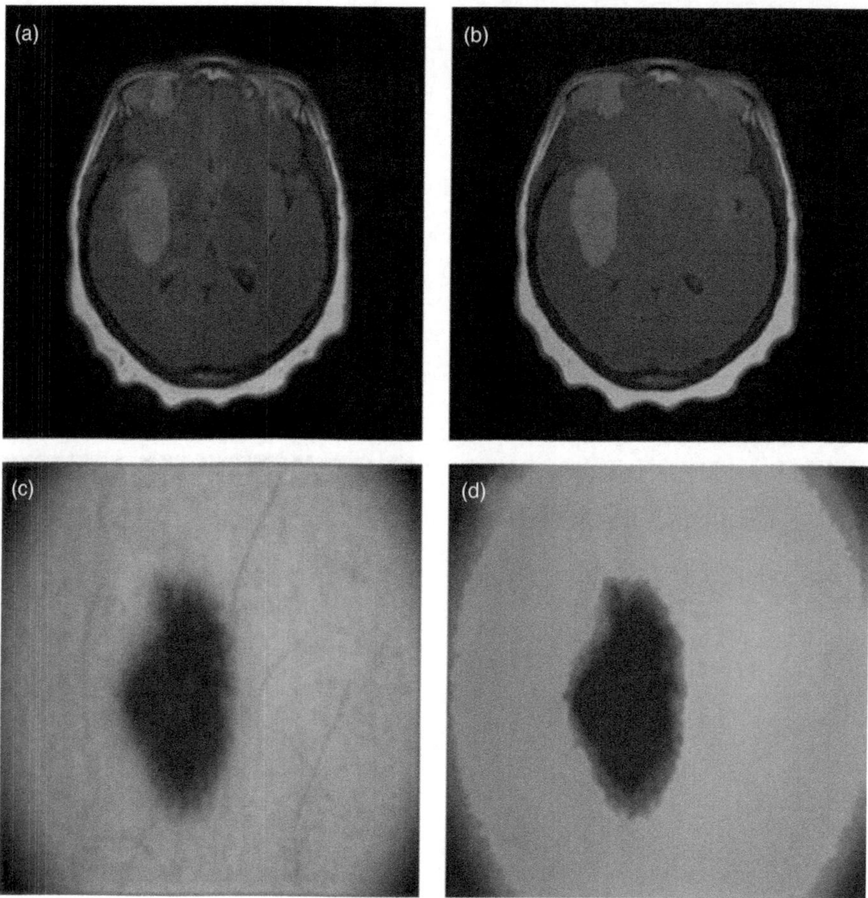

FIGURE 1.3 Results of region-based segmentation: (a) and (c) are original brain tumor and skin cancer images, (b) and (d) are segmented results of brain and skin cancer images.

where X denotes the various regions in the image I. X_a, X_b, $X_{c......}X_z$ are the regions which possess common characteristics. The region-based algorithms look for the homogenous characteristics of regions for the segmentation task. The results of region-based segmentation are shown in Figure 1.3. Region-based algorithms are implemented in the following steps:

a) **Extraction of Features**: Extracting pertinent features from the image, such as color, intensity, texture, or shape, is the first step in the process. The various regions of the image will be distinguished using these features.
b) **Seed Selection**: Choosing the initial seed points or regions is a common first step in region-based segmentation. These seeds can be selected automatically or by hand according to predetermined standards. Seeds are the beginning of a region's growth or fusion.

c) **Region Growing or Splitting**: There are two typical methods for segmenting data based on regions:
   - **Region Growing**: This method begins with seed points and, if nearby pixels share the selected feature(s), adds them to the region iteratively. Until a halting condition is satisfied, this process keeps going.
   - **Region Splitting**: This is also known as hierarchical segmentation, is a technique that starts with the entire image as a single region and divides it into smaller regions recursively according to how dissimilar the feature(s) within the region are. A structure of regions resembling a tree may arise from this hierarchical segmentation.

d) **Merging and Post-processing**: To obtain the required segmentation, regions may need to be merged or refined after expanding or splitting. Small or undesirable regions can be eliminated and the final segmentation result can be improved by using post-processing techniques.

e) **Labeling**: Provide the segmented regions labels so that they can be further examined or used for object recognition.

1.2.4 Split and Merge Algorithm

An image can be segmented using the Split and Merge algorithm, which separates an image into segments or regions according to predetermined criteria (An et al. 2015). It works especially well for segmenting images that have areas with uniform texture, color, or other attributes. The following steps constitute split and merge algorithm:

1. **Initialization**: The image is initially seen as a single region.
2. **Splitting**:
   - **Splitting Criteria**: Assess the area to see if it can be divided into more manageable sections. Usually, there are some criteria used to determine this, like variations in the region's color, intensity, or texture.
   - **Splitting Process**: Divide the region into smaller sub-regions if the requirements for splitting are satisfied. The region can be divided into four equal quadrants to achieve this, or more complex methods such as quadtree or octree decomposition can be used.
3. **Merging**:
   - **Merging Criteria**: After splitting, determine whether the resulting sub-regions should be merged. Examining the degree of similarity between adjacent sub-regions is frequently one of the merging criteria. Sub-regions are merged back together if they are sufficiently similar to one another.
   - **Procedure for Merging Regions**: Adjacent regions that comply with the merging criteria are combined. By taking this step, it is ensured that the segmentation is not too finely tuned.
4. **Iterative Process**: Until no more splitting or merging is possible, steps 2 and 3 are repeated several times. A collection of regions that represent interesting objects or features within the image is the end product.

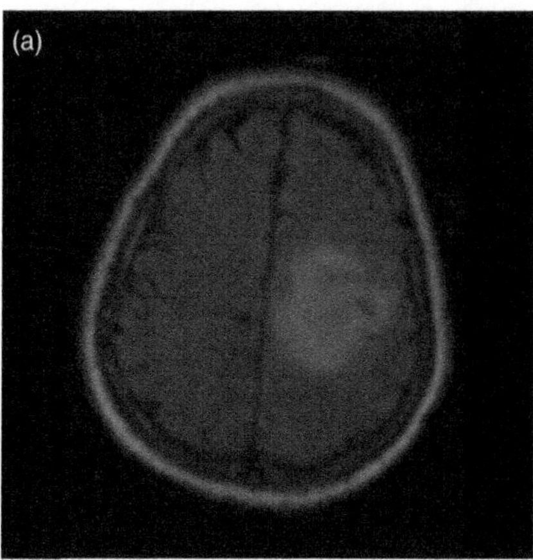

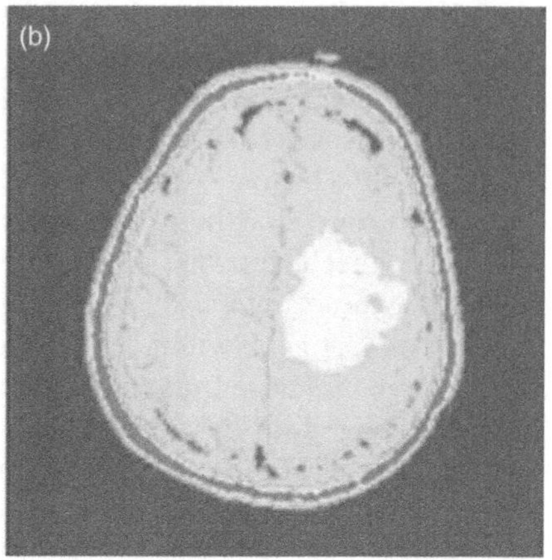

FIGURE 1.4 Results of split and merge algorithm-based segmentation: (a) original brain tumor image and (b) segmented results of brain images.

5. **Post-processing**: To further refine the segmented regions, additional post-processing steps may be carried out, depending on the application. This could involve other improvements, such as boundary smoothing or noise reduction.

The Split and Merge algorithm can be applied to a variety of image segmentation tasks because it is versatile. This method, which is recursive in nature and divides an image into homogeneous regions, is helpful for tasks like edge detection, object recognition, and scene analysis. By selecting suitable split and merge criteria based on the unique properties of the processed images, the algorithm can be tailored. The results after applying the split and merge algorithm on brain MR images can be seen in Figure 1.4.

1.3 ADVANCED DEEP LEARNING-BASED METHODS FOR IMAGE SEGMENTATION

In the field of medical image segmentation, convolutional neural networks, or CNNs, have indeed become a disruptive technology (Krizhevsky, Sutskever, and Hinton 2017). Anatomical structures and lesions in medical images, including X-rays, CT scans, MRIs, and histopathological slides, can be automatically and accurately segmented by CNNs, a remarkable capability. Human error and variability in the interpretation of medical images could be decreased by this technology. In remote or underserved areas where access to medical specialists is limited, CNNs can be used to provide specialized medical expertise. Rapid medical image processing by CNNs enables more rapid treatment planning and diagnosis. Early identification of illnesses and abnormalities can be detected by CNNs, which can lead to better patient outcomes through early intervention (Badrinarayanan, Kendall, and Cipolla 2016). The exploration of advanced deep learning for image segmentation continues to push the envelope of what is possible and has the potential to completely transform how humans view and work with visual data. The U-Net and GAN deep learning-based architectures will be covered in more detail in this section.

1.3.1 U-Net

In deep learning, U-Net (Ronneberger, Fischer, and Brox 2015) is a popular architecture for segmenting images. Olaf Ronneberger, Philipp Fischer, and Thomas Brox first described it in their paper "U-Net: Convolutional Networks for Biomedical Image Segmentation," published in 2015. U-Net has gained popularity in medical image analysis and is especially well-suited for tasks involving semantic and instance segmentation. The combination of an encoder (contracting path) and a decoder (expansive path) makes up the U-shaped structure of the U-Net architecture as shown in Figure 1.5. For tasks involving semantic segmentation, this model performs exceptionally well. The encoder is comprised of two consecutive 3 × 3 convolutions (conv) with zero padding and a rectified linear unit (ReLU) (Agarap 2019) activation function. To connect levels or perform down sampling, a max-pooling operation with stride 2 is employed. At every level after that, the channel number of feature maps is doubled. A 2 × 2 up-convolution (up-conv) is used in the symmetric decoder counterpart in addition to up sampling, in order to cut the number of channels in half. To maintain the low-level information, the center-cropped feature map from the encoder is sent via skip connections to the decoder at each level. For concatenation, cropping is required to keep feature maps of the same size. We then apply ReLU and two more 3 × 3 convolutions. The channel number is finally converted to the required number of classes C using a 1 × 1 convolution. The network creates a segmentation map with C classes in this configuration based on a 2D image as input. Segmentations performed by U-Net are shown in Figure 1.6.

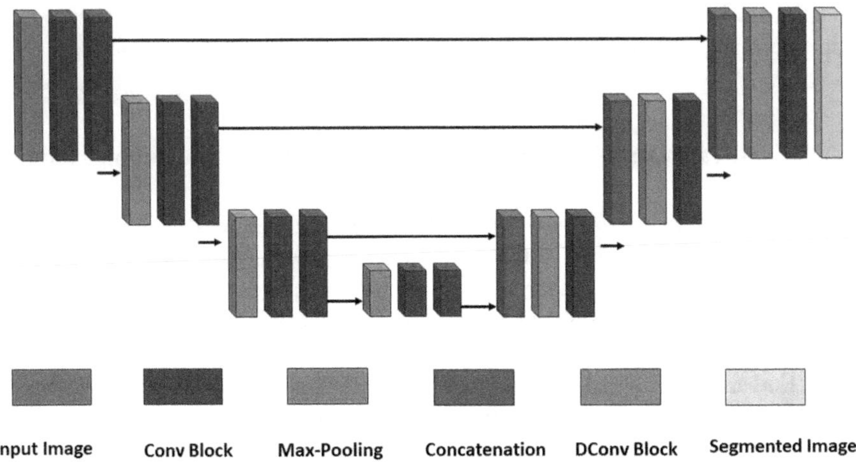

FIGURE 1.5 U-Net Architecture.

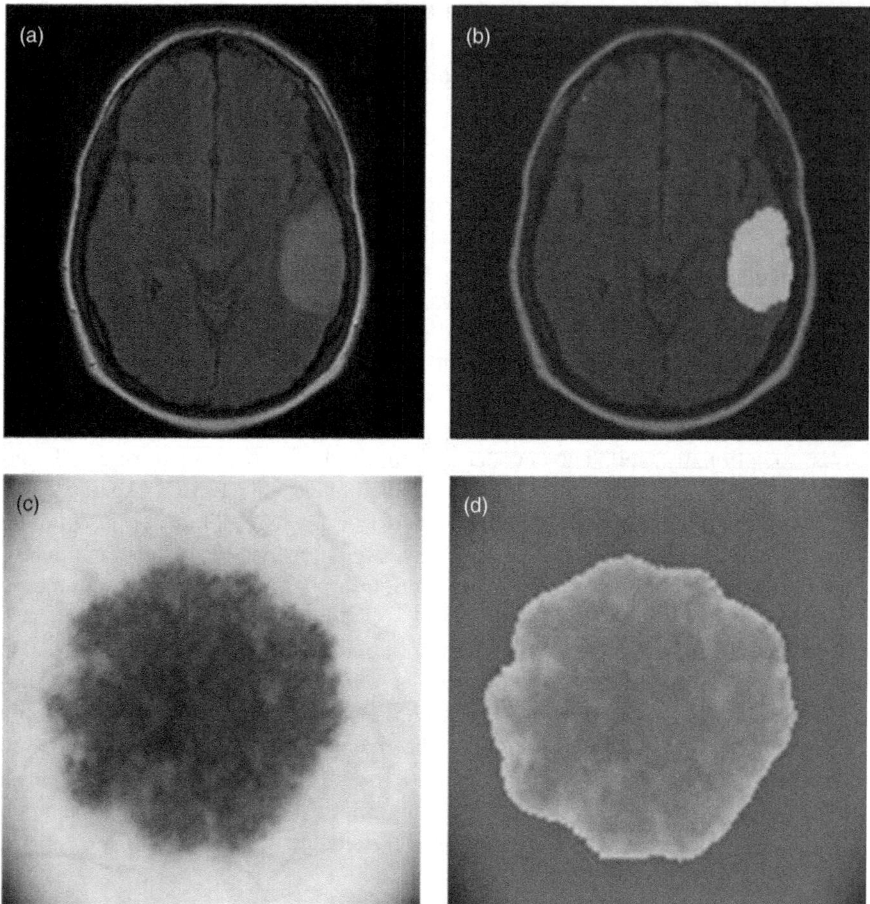

FIGURE 1.6 Results of U-Net-based segmentation (a) and (c) are original brain tumor and skin cancer images, (b) and (d) are segmented results of brain and skin cancer images.

1.3.1.1 Advantages of U-Net over Traditional Image Segmentation Methods

1. **Excellent Precision and Adaptability**: U-Net has proven to be highly accurate and performs at the cutting edge in a variety of image segmentation tasks, especially in the analysis of medical images. It is a flexible option for various applications because it can be adjusted to a broad variety of segmentation tasks.
2. **End-to-End Learning**: Feature descriptions are directly learned by U-Net from the input data. It can adjust to the unique properties of the data because it doesn't rely on manually constructed feature engineering.
3. **Sturdy against Variability**: Conventional techniques frequently depend on manually adjusted parameters and presumptions about the properties of the

images, which might not always be accurate. Because U-Net is data-driven, it is more resistant to changes in the data.

4. **Retrieves Local and Contextual Data**: Through its encoder-decoder architecture, U-Net is able to capture both local and contextual information, which facilitates precise and context-aware segmentations.
5. **Maintaining Spatial Information**: Fine-grained spatial information is preserved in many segmentation tasks by U-Net's use of skip connections between the encoder and decoder layers.
6. **Decreased Manual Annotation Work**: Conventional segmentation techniques frequently necessitate laborious manual feature engineering and training annotation efforts. Through transfer learning, U-Net can lessen the need for manually created features and, in certain situations, the effort required for annotation.
7. **Automation**: Utilizing U-Net can reduce the need for human operators to complete labor-intensive and complex segmentation tasks by enabling fully automated segmentation tasks.
8. **Progressive Training**: The network can learn gradually from coarse features to fine details due to U-Net's encoder-decoder architecture, which is crucial for tasks like medical image segmentation.
9. **Ability to Scale and Adapt**: Widely applicable to a variety of image segmentation problems, U-Net models can be trained on large datasets and exhibit good generalization to new data.

In complex and varied segmentation tasks in particular, U-Net proves beneficial when high accuracy, generalization, and automation are needed. In situations where interpretability, computational efficiency, and domain-specific knowledge are critical, traditional approaches might still be useful. The specific requirements of the application, the availability of data, and the available computational resources all influence the decision between U-Net and more conventional approaches. In order to get the best of both worlds, hybrid approaches—which combine the advantages of both U-Net and conventional methods—are frequently investigated.

1.3.2 Generative Adversarial Network

In 2014, Ian Goodfellow and associates developed a class of machine learning algorithms called Generative Adversarial Networks or GANs (Goodfellow et al. 2014). The purpose of GANs is to create fresh, never-before-seen data that closely resembles a given set of training data. A generator and a discriminator are two neural networks that interact in the fundamental idea of GANs. Figure 1.7 shows the overall architecture of GAN. The generator is a neural network that attempts to produce data samples (such as images) that resemble the training data by using random noise or random input as its input.

The generator generates random noise at first, and the samples it creates are usually meaningless and dissimilar to the actual data. An additional neural network that assesses the veracity of a particular data sample is called the discriminator. It distinguishes between real and false input by using samples of created data from the

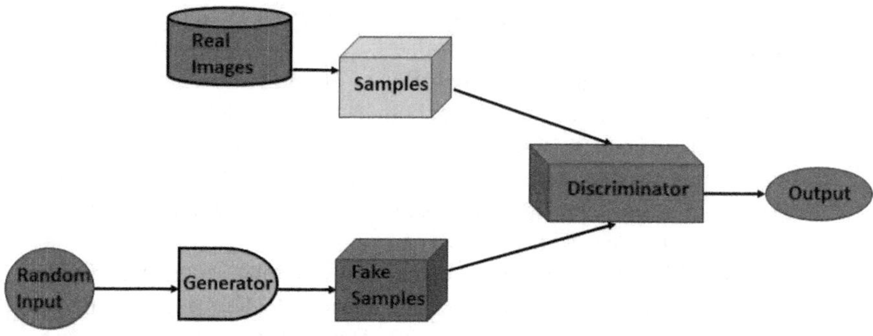

FIGURE 1.7 GAN Architecture.

generator and real data samples from the training set. The discriminator is taught to discern between authentic and counterfeit samples. Since the generated samples are random and don't accurately reflect the real data, it performs poorly at first.

These two networks engage in competitive interaction throughout the GAN's training phase.

a) **Training the Discriminator**
The discriminator is trained using samples of created data from the generator labeled "fake" and real data samples labeled "real." The objective of the discriminator is to accurately identify phony and authentic samples. With time, it learns to become more accurate.
b) **Training the Generator**
The generator creates samples by using random noise as input. The discriminator receives these created samples, and the generator's goal is to create samples that it deems to be "real." Enhancing the generator's capacity to produce samples that are identical to genuine data is the goal.
c) **Adversarial Training**
In an adversarial way, the discriminator and generator are trained concurrently. The discriminator improves its ability to discern between authentic and phony samples as the generator becomes more adept at producing realistic samples. Both networks continuously improve their performance as a result of this adversarial process.

In the context of image or data segmentation, a GAN is used to create segmented output from an input image or set of data. The generator, or part of the GAN, creates the segmented images, and the discriminator, or other part of the GAN, assesses the precision and quality of the segmentation. The segmentation results are more accurate and of higher quality thanks to this adversarial approach. GANs have been applied to a variety of segmentation tasks, including object segmentation in computer vision and medical image segmentation.

The training process keeps going until a predetermined threshold is reached, like a predetermined number of iterations or the generated samples attain the required

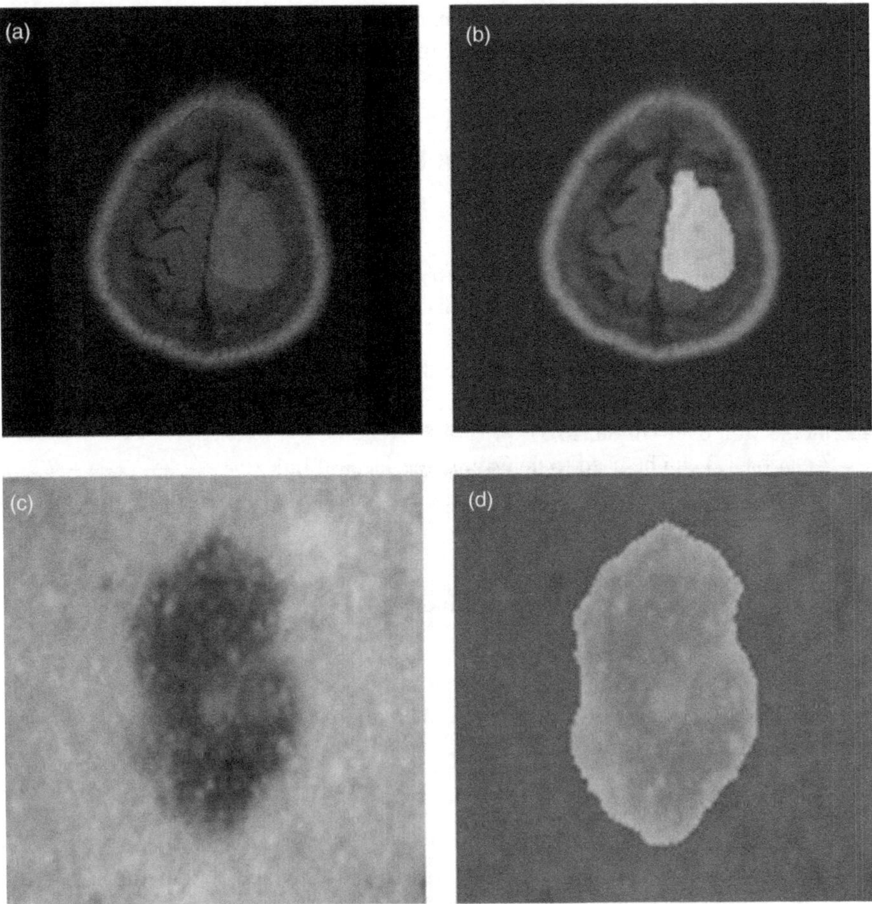

FIGURE 1.8 Results of GAN-based segmentation: (a) and (c) are original brain tumor and skin cancer images, (b) and (d) are segmented results of brain and skin cancer images.

level of realism. In a perfect world, the discriminator is unable to consistently discern between created and genuine samples, and the generator learns to produce examples that are comparable to the training data. The segmented results obtained by GAN network on brain and skin cancer images are shown in Figure 1.8.

1.3.2.1 Advantages of GAN over Traditional Methods

Regarding circumstances like uncommon environmental conditions or rare medical disorders, when real data collection is challenging, GANs can be used to create synthetic data. To help segmentation models handle boundary instances, this can be used during training. To improve the delineation of object borders, GANs can be employed in conjunction with conventional segmentation approaches. In order to enable segmentation models provide more accurate and precise findings, GANs can produce images with more clear object boundaries. High-level semantic features can be learned by GANs from the data. Semantic information can be added to the segmentation process by employing GANs in a multi-modal manner, which improves the segmentation outcomes.

1.4 EVALUATION METRICS FOR IMAGE SEGMENTATION

a) **Jaccard Index**

A metric for comparing two sets' similarity is the Jaccard Index, sometimes referred to as the Jaccard Similarity Coefficient (Everingham et al. 2015). It is frequently utilized in many different domains, including data mining, natural language processing, and information retrieval. By comparing the intersection and union of two sets, the Jaccard Index calculates how similar the two sets are to one another. It is also known as Intersection over Union (IoU).

Equation 1.3 can be used to determine the Jaccard Index:

Jaccard Index (J) = (Size of Intersection of Sets A and B)/(Size of Union of Sets A and B)

$$\text{or, Jacquard Index} = \frac{S \bigcap T}{S \bigcup T} \tag{1.3}$$

where T is the ground truth data and S is the anticipated segmentation data. A result between 0 and 1, where 0 indicates no segmentation and 1 indicates an accurate segmentation, is what Jaccard Index returns.

b) **Dice Score**

The DICE score (Martin, Fowlkes, and Malik 2004), which is sometimes referred to as the Sørensen–Dice coefficient or Dice similarity coefficient, is a frequently employed metric in image segmentation that is used to assess the precision of a segmentation technique or model quantitatively. It calculates how close an image's expected and ground truth segmentations are to each other. The DICE score is a number between 0 and 1, where 0 denotes no overlap and 1 denotes a perfect match between the expected and ground truth segmentations. Equation 1.4 denotes the mathematical expression for dice score:

$$\text{Dice Score} = \frac{2 \times S \bigcap T}{S + T} \tag{1.4}$$

In order to provide a balanced measure of segmentation quality, the DICE score considers both the true positive (properly segmented) and false positive (over-segmented) and false negative (under-segmented) regions. This makes it a useful metric in the evaluation of image segmentation. Greater accuracy in segmentation is indicated by higher DICE values, while lower scores imply less accurate segmentations. To thoroughly evaluate the effectiveness of segmentation models, it is one of several metrics that are combined with others, such as Intersection over Union.

c) **Precision**

The percentage of correctly categorized positive pixels (such as accurately segmented object pixels) among all the positive pixels predicted by the model is referred to as precision in image segmentation (Villmann et al. 2014). It gauges how well the model predicts things. Precision can be expressed by Equation 1.5:

$$\text{Precision} = \frac{\text{TP}}{\text{TP} + \text{FP}} \quad (1.5)$$

d) **Recall**

The ratio of true positive pixels to all actual positive pixels in the ground truth segmentation is known as recall in image segmentation (Taha and Hanbury 2015). It gauges how well the model can recognize every real positive pixel. Equation 1.6 mathematically expresses the recall metric:

$$\text{Recall} = \frac{\text{TP}}{\text{TP} + \text{FN}} \quad (1.6)$$

False positives (FP) are pixels that are incorrectly classified as being a part of the target item, whereas true positives (TP) are pixels that are accurately identified as belonging to the target object. Pixels in the ground truth that the model failed to detect are known as false negatives (FN).

1.5 ADVANCED AI (DEEP LEARNING) AS A DISRUPTIVE TECHNOLOGY IN MEDICAL IMAGE SEGMENTATION

Medical image segmentation has found deep learning to be a very disruptive technique. Medical image segmentation, which entails locating and defining anatomical structures, tumors, lesions, and other regions of interest within medical images like CT scans, MRIs, X-rays, and histopathology slides, is an essential task in healthcare and clinical diagnosis. Diagnostic image segmentation has shown impressive accuracy when using deep learning models, particularly convolutional neural networks (CNNs) and their derivatives. They can reliably and automatically identify and segment structures that, because of minute fluctuations or noise in the images, could be difficult for human experts to identify. Semantic segmentation (classifying structures into groups such as organs or tumors) and instance segmentation (identifying distinct instances of the same structure) are made possible by deep learning models (Avendi, Kheradvar, and Jafarkhani 2015). This is very helpful for tracing the evolution of particular nodules over time or measuring the number of lesions. Models with relatively small labeled datasets can be adapted to new medical image segmentation tasks through transfer learning and pre-trained models. This lessens the need for manual annotation, which is expensive and time-consuming. Through the identification of tiny abnormalities in medical imaging, deep learning models can aid in the early detection of diseases like cancer or neurodegenerative disorders (Bernard et al. 2018). This could lead to early treatments and better patient outcomes. Deep learning models can be tailored to certain patient demographics and medical situations. This flexibility is especially helpful in areas with varying requirements and standards, such as cardiology, cancer, and radiology.

Deep learning has made a strong technological statement over more conventional approaches in a variety of fields, including computer vision, natural language processing, and medical imaging. Feature learning is one of its main advantages since deep learning models can automatically extract complex features from data, eliminating the need for human feature engineering (Luo et al. 2020). Not only is this automatic feature learning effective, but it also improves generalization over a variety of datasets. The scalability and adaptability of deep learning are also attractive features. Large data volumes and processing resources can be easily accommodated by it, which makes it appropriate for tasks involving substantial data growth. Deep learning models are remarkably flexible because they can be tailored for many applications using a same architecture, like a recurrent neural network (RNN) or convolutional neural network (CNN), which eliminates the need for task-specific methods.

Deep learning has some challenges despite its amazing benefits. For training, it usually requires large volumes of labeled data, which might not be available for every task. Large amounts of calculation may be required, requiring expensive, powerful technology. Deep learning models' interpretability is a persistent problem, particularly in delicate industries like finance and healthcare. One potential issue that calls for the use of regularization approaches is overfitting. When implementing deep learning, ethical issues with prejudice, justice, and privacy must also be properly addressed. In general, although deep learning has notable benefits over conventional techniques, its implementation must be approached mindful of its constraints and moral ramifications. It's a powerful technology that can spur innovation in many different businesses and transform many other fields when used carefully.

REFERENCES

Abler, Daniel, Russell C. Rockne, and Philippe Büchler. 2019. "Evaluating the Effect of Tissue Anisotropy on Brain Tumor Growth Using a Mechanically Coupled Reaction–Diffusion Model." In *New Developments on Computational Methods and Imaging in Biomechanics and Biomedical Engineering*, edited by João Manuel R. S. Tavares and Paulo Rui Fernandes, 33:37–48. Cham: Springer International Publishing. https://doi.org/10.1007/978-3-030-23073-9_3

Agarap, Abien Fred. 2019. "Deep Learning Using Rectified Linear Units (ReLU)." *arXiv:1803.08375 [Cs, Stat]*, February. http://arxiv.org/abs/1803.08375

An, Hongsub, Hyeon-min Shim, Sang-il Na, and Sangmin Lee. 2015. "Split and Merge Algorithm for Deep Learning and Its Application for Additional Classes." *Pattern Recognition Letters* 65 (November): 137–144. https://doi.org/10.1016/j.patrec.2015.07.024

Avendi, M. R., A. Kheradvar, and H. Jafarkhani. 2015. "A Combined Deep-Learning and Deformable-Model Approach to Fully Automatic Segmentation of the Left Ventricle in Cardiac MRI." *arXiv:1512.07951 [Cs]*, December. http://arxiv.org/abs/1512.07951

Badrinarayanan, Vijay, Alex Kendall, and Roberto Cipolla. 2016. "SegNet: A Deep Convolutional Encoder-Decoder Architecture for Image Segmentation." *arXiv:1511.00561 [Cs]*, October. http://arxiv.org/abs/1511.00561

Balafar, M. A., A. R. Ramli, M. I. Saripan, and S. Mashohor. 2010. "Review of Brain MRI Image Segmentation Methods." *Artificial Intelligence Review* 33 (3): 261–274. https://doi.org/10.1007/s10462-010-9155-0

Bauer, Stefan, Roland Wiest, Lutz-P Nolte, and Mauricio Reyes. 2013. "A Survey of MRI-Based Medical Image Analysis for Brain Tumor Studies." *Physics in Medicine and Biology* 58 (13): R97–129. https://doi.org/10.1088/0031-9155/58/13/R97

Bernard, Olivier, Alain Lalande, Clement Zotti, Frederick Cervenansky, Xin Yang, Pheng-Ann Heng, Irem Cetin, et al. 2018. "Deep Learning Techniques for Automatic MRI Cardiac Multi-Structures Segmentation and Diagnosis: Is the Problem Solved?" *IEEE Transactions on Medical Imaging* 37 (11): 12.

Caraiman, Simona, and Vasile I. Manta. 2014. "Histogram-Based Segmentation of Quantum Images." *Theoretical Computer Science* 529 (April): 46–60. https://doi.org/10.1016/j.tcs.2013.08.005

Everingham, Mark, S. M. Ali Eslami, Luc Van Gool, Christopher K. I. Williams, John Winn, and Andrew Zisserman. 2015. "The Pascal Visual Object Classes Challenge: A Retrospective." *International Journal of Computer Vision* 111 (1): 98–136. https://doi.org/10.1007/s11263-014-0733-5

Goodfellow, Ian J., Jean Pouget-Abadie, Mehdi Mirza, Bing Xu, David Warde-Farley, Sherjil Ozair, Aaron Courville, and Yoshua Bengio. 2014. "Generative Adversarial Networks." *arXiv:1406.2661 [Cs, Stat]*, June. http://arxiv.org/abs/1406.2661

Krizhevsky, Alex, Ilya Sutskever, and Geoffrey E. Hinton. 2017. "ImageNet Classification with Deep Convolutional Neural Networks." *Communications of the ACM* 60 (6): 84–90. https://doi.org/10.1145/3065386

Liu, Xiangbin, Liping Song, Shuai Liu, and Yudong Zhang. 2021. "A Review of Deep-Learning-Based Medical Image Segmentation Methods." *Sustainability* 13 (3): 1224. https://doi.org/10.3390/su13031224

Luo, Qianhui, Huifang Ma, Li Tang, Yue Wang, and Rong Xiong. 2020. "3D-SSD: Learning Hierarchical Features from RGB-D Images for Amodal 3D Object Detection." *Neurocomputing* 378 (February): 364–374. https://doi.org/10.1016/j.neucom.2019.10.025

Martin, D.R., C.C. Fowlkes, and J. Malik. 2004. "Learning to Detect Natural Image Boundaries Using Local Brightness, Color, and Texture Cues." *IEEE Transactions on Pattern Analysis and Machine Intelligence* 26 (5): 530–549. https://doi.org/10.1109/TPAMI.2004.1273918

Ronneberger, Olaf, Philipp Fischer, and Thomas Brox. 2015. "U-Net: Convolutional Networks for Biomedical Image Segmentation." *arXiv:1505.04597 [Cs]*, May. http://arxiv.org/abs/1505.04597

Taha, Abdel Aziz, and Allan Hanbury. 2015. "Metrics for Evaluating 3D Medical Image Segmentation: Analysis, Selection, and Tool." *BMC Medical Imaging* 15 (1): 29. https://doi.org/10.1186/s12880-015-0068-x

Tripathi, Sumit, Taresh Sarvesh Sharan, Shiru Sharma, and Neeraj Sharma. 2021. "Segmentation of Brain Tumour in MR Images Using Modified Deep Learning Network." In *2021 8th International Conference on Smart Computing and Communications (ICSCC)*, 1–5. Kochi, Kerala, India: IEEE. https://doi.org/10.1109/ICSCC51209.2021.9528298

Tripathi, Sumit, and Neeraj Sharma. 2021. "Computer-Based Segmentation of Cancerous Tissues in Biomedical Images Using Enhanced Deep Learning Model." *IETE Technical Review*, November, 1–15. https://doi.org/10.1080/02564602.2021.1994044

Villmann, Thomas, Marika Kaden, Mandy Lange, Paul Sturmer, and Wieland Hermann. 2014. "Precision-Recall-Optimization in Learning Vector Quantization Classifiers for Improved Medical Classification Systems." In *2014 IEEE Symposium on Computational Intelligence and Data Mining (CIDM)*, 71–77. Orlando, FL, USA: IEEE. https://doi.org/10.1109/CIDM.2014.7008150

Xu, Ziqi, Xiaoqiang Ji, Meijiao Wang, and Xiaobing Sun. 2021. "Edge Detection Algorithm of Medical Image Based on Canny Operator." *Journal of Physics: Conference Series* 1955 (1): 012080. https://doi.org/10.1088/1742-6596/1955/1/012080

2 Applications of Transformer in Medical Imaging
A Review

Satish Kumar
BGSB University Rajouri, (J&K), India

2.1 INTRODUCTION

Transformers have emerged as the potential approach in Natural Language Processing (NLP) which have major applications in several domains such as language generation, synthesis, speech recognition, and text-to-audio conversion (Shamshad et al. 2023). Several Transformer methods and models, trained on extensive datasets, have gained popularity in addressing natural language processing tasks that include new language generation tool called Bidirectional Encoder Representations from Transformers (BERT) (Brown et al. 2020).

The potential of deep convolutional neural networks and their variants in various computer vision tasks is attributed to understand image representation via hierarchical parsing, capturing vision semantics, which is critical for successful generation of computer vision models (Krizhevsky et al. 2012). Despite their success, convolutional neural networks have a limitation to ignore long-range dependencies that is preset inside image. Drawing inspiration from Transformer's success in NLP introduced the vision-based Transformers. They successfully addressed the long-term dependencies within the input image by using image formulation as a series of prediction challenges for the image patch area. Transformer and its modifications have shown promise across a variety of benchmark datasets. Transformers are commonly used in computer vision applications, such as in image categorization, detection, segmentation, and synthesis.

Transformers have recently made significant contributions in medical image analysis, finding applications in disease diagnosis (Zhang and Wen 2021) and various clinical purposes. Figure 2.1 illustrates Transformers evolution in the medical image analysis, highlighted, methodologies pertaining to categorization, segmentation, detection, and synthesis applications.

The chapter includes comprehensive discussions on the development of Transformer-based approaches for solving complicated real-world challenges, including weakly supervised, multi-task, and multi-modal learning paradigms.

DOI: 10.1201/9781032644509-2

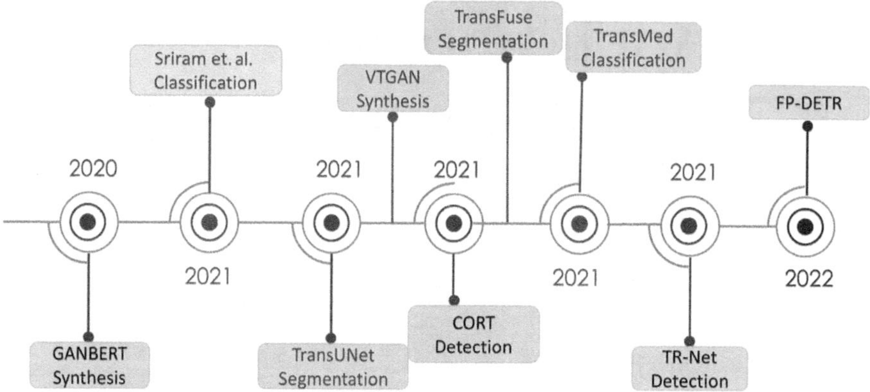

FIGURE 2.1 Transformers evolution in the medical image analysis, highlighted, methodologies pertaining to classification, detection, segmentation, and synthesis applications.

The organization of the chapter is arranged as follows: Section 2.2 explains the fundamentals of Transformers. Section 2.3 presents a summary of current Transformer uses in medical image processing, while Section 2.5 digs into Transformer potential applications. The concluding thoughts are offered in Section 2.8.

2.2 TRANSFORMERS

In typical Transformer architecture, the attention mechanism within neural networks plays an important role. Therefore, we begin this section by explaining the basic idea behind the attention mechanism, then explore in extensive detail about how the Transformer bases models in computer vision work.

2.2.1 Attention Mechanism

When it comes to information exploration, humans typically use their "attention mechanism" to find out meaningless context and focus on the important details they come across every day. In the field of deep learning, researchers have created attention mechanisms based on this observation. These mechanisms are particularly good at sorting through homogeneous data and focusing on the most important parts or constituents. Bahdanau et al. (2014) first proposed a language translation task, which computes the weighted sum of all annotations and improves the translation performance using encoder-decoder model.

2.2.2 Attention Mechanism for Computer Vision

Similar principles of attention mechanisms have been applied to computer vision. Hu et al. (2020), for example, proposed an innovative attention mechanism, Squeeze-and-Excitation, that was designed to perform feature re-calibration. This method prioritizes informative characteristics for a given visual task while downplaying other aspects.

Self-attention: To speed up parallel calculations, the self-attention process is usually performed as a matrix. To quickly demonstrate the idea of self-attention, we will first characterize it in terms of its constituent elements. Learnable weight matrices, such as Queries (Q), Keys (K), and Values (V), are critical components of the self-attention process. These matrices are critical in assessing the relative relevance of each patch to the overall picture.

The input sequence, represented as matrix X, undergoes Transformation through the weight matrices to generate Query (Q), Key (K), and Value (V) matrices. This Transformation serves as the foundation for calculating the attention matrix, which captures the interdependencies among different patches. The attention matrix is calculated by applying the softmax function on the scaled dot product of Queries and Keys. This stage enables the model to give weights to various patches depending on their relationship to one another. The final output of the self-attention layer, denoted as Z, is obtained by multiplying the attention matrix (A) with the Values matrix (V). This step consolidates the weighted information and produces a refined representation of the input sequence.

2.3 ARCHITECTURE

Vaswani et al. (2017) introduced a conventional Transformer architecture that contained an encoder-decoder model in their work. The encoder Transforms an input sequence $x_1, \ldots, x_n$ into an output sequence $z_1, \ldots, z_n$ of equal length. The decoder creates the output $y_1, \ldots, y_m$ element-by-element using the encoded representation z, taking the previous output as an extra input. Figure 2.2 depicts a typical Transformer architecture, which is discussed in the following section.

2.3.1 Encoder

In a Transformer, the encoder is made up of six stacked blocks with two sorts of layers: multi-head attention and feed-forward. These layers are modified with layer normalization and residual connections. Each block computes the multi-head attention first, followed by layer-wise normalization, which calculates the total of the multi-head attention's input and output. The sum of the feed-forward layer's input and output is then normalized layer-wise.

2.3.2 Decoder

The decoder, similar to the encoder, is made up of six blocks, with a few small changes. An extra layer of self-attention is added to the encoded output. Given that the prediction is based on a known state, the initial self-attention layer employs masking to prevent further contributions to the prior position's state. A linear layer and a softmax layer are added at the decoder's end.

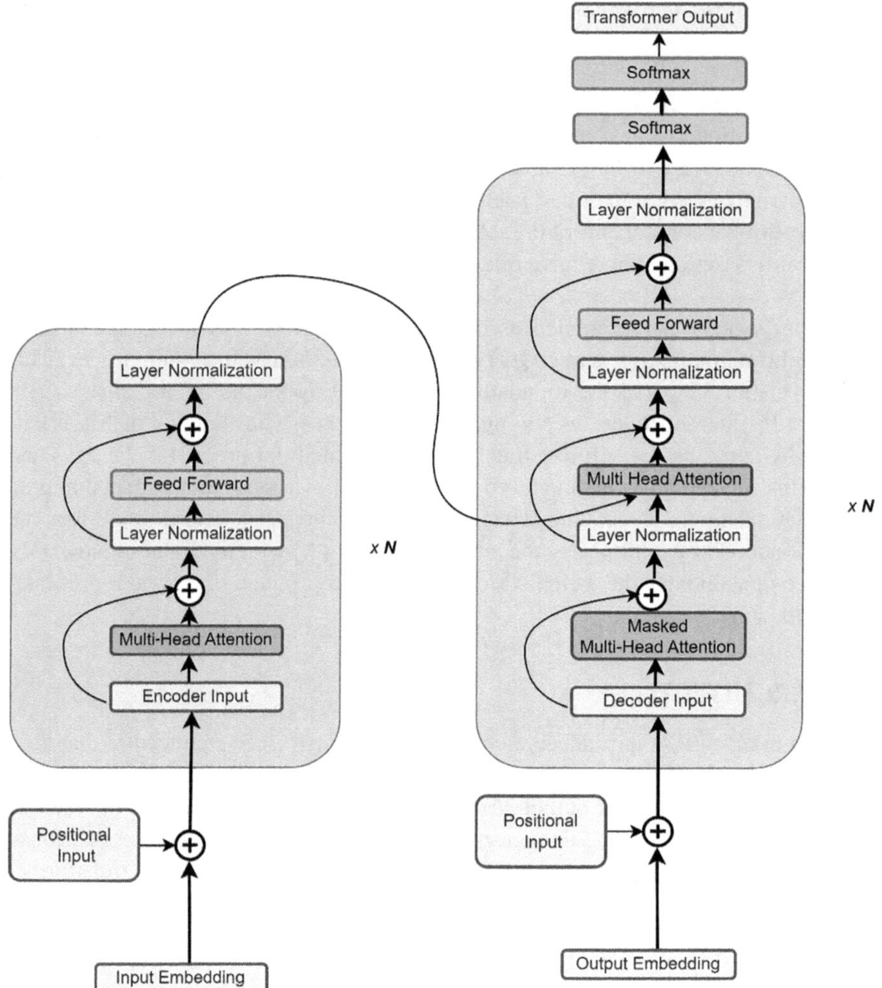

FIGURE 2.2 Demonstration standard Transformer architecture (Chen et al. 2021b).

2.4 VISION-BASED TRANSFORMERS

Transformers' success in Natural Language Processing has had a considerable influence on the Computer Vision research community, driving a number of initiatives to apply Transformers to vision problems. Transformer architectures in computer vision have advanced rapidly, with significant examples such as Vision-based Transformer (Dosovitskiy et al. 2021 and many others).

ViT: The following Transformer serves as an image categorization model, following the basic design of a regular Transformer. A Transformer converts the input image into a sequence of patches, each using a positional encoding algorithm that encodes spatial locations. Several patches, together with a class token, are then fed into the Transformer, which calculates Multi-Head Self-Attention and outputs the patches' learnt embedding.

Swin Transformer: Liu et al. (2021) proposed the Swin Transformer, which addresses the computational constraints involved with processing high-resolution photos and handling varied patch sizes for scene interpretation tasks. Their novel technique included the use of window self-attention to reduce computing complexity. They also employed shifting window attention to efficiently capture and simulate cross-window connections, which improved the system's overall efficiency.

2.5 APPLICATIONS OF TRANSFORMER IN MEDICAL IMAGE

Transformers have found extensive application in comprehensive clinical solutions. This segment begins by presenting applications of Transformer-based medical image analysis, encompassing tasks such as classification, segmentation, detection, and synthesis. Figure 2.3 shows Transformers applications in medical image analysis.

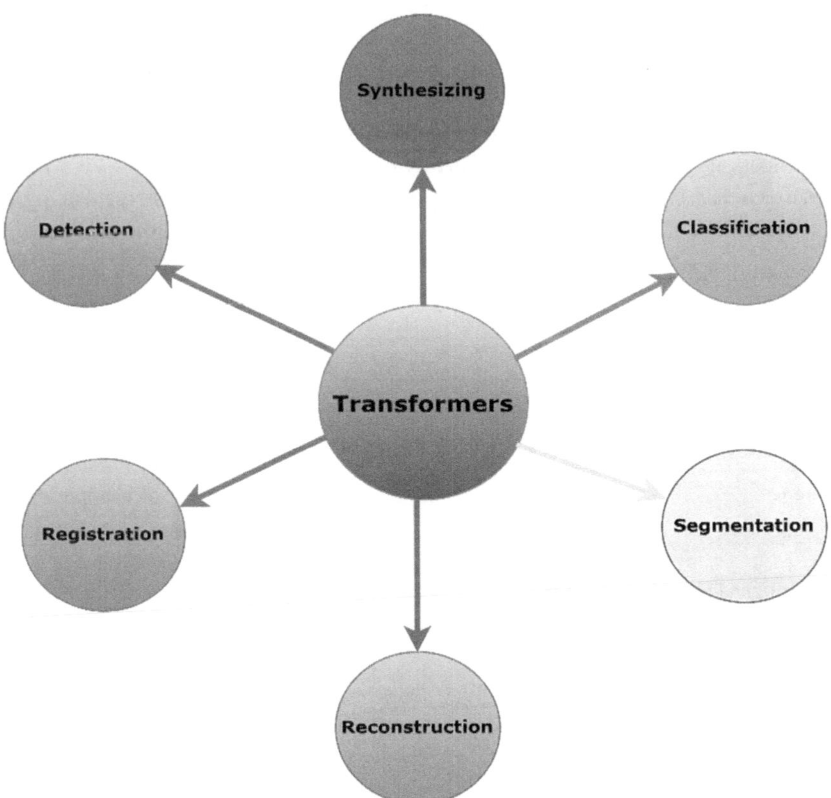

FIGURE 2.3 Transformers applications in medical image analysis.

2.5.1 Classification

The classification tasks are further detailed in this section, where methods utilizing Transformers for disease diagnosis are categorized into three classes:

1. Direct application of Vision Transformers to medical images,
2. Integration of Transformers with convolutions for enhanced local features learning,
3. Fusion of Transformers with graph representations to effectively manage intricate data structures.

The classification tasks on medical images are systematically explored using Transformer summarized in Table 2.1.

2.5.1.1 Pure Transformers in Classification

Transformers that closely resemble the originally proposed Vision Transformer (Dosovitskiy et al. 2021) are termed "pure Transformers." These approaches typically maintain a structure similar to the original method. The literature on pure Transformers is presented based on different image modalities.

2.5.1.2 Hybrid Transformer in Classification

While pure Vision Transformers can generate promising results with modest adjustments, extensive research has been conducted to investigate the possible benefits of merging Transformers using other learning features with the goal of increasing the collection of complicated data distributions or achieving improved performance. Transformers are frequently used in conjunction with convolutional layers and graph representations, and both types are covered here.

TABLE 2.1
Applications of Transformers in Classification of Medical Images

Reference	Method	Category	Medical Application
(Liu et al. 2022)	Multi Scale	Cytopathology	Cervical cancer classification
(Dai, Gao, and Liu 2021)	TransMed	MRI	Multi-modal classification: disease classification, lesion identification
(Zhang et al. 2022)	TractoFormer	Diffusion MRI	Fiber representation and nerve tracts modelling
(Lyu et al. 2022)	Brain metastases classification nervous system	MRI	Classification of the brain tumor of central
(Stegmuller et al. 2023)	ScoreNet	Histology Datasets	Classification of breast cancer

***Transformers with Convolutions*:** Transformers are more concerned with modeling global relationships within data, whereas traditional Convolutional Neural Networks are more concerned with local textures. Recognizing these distinctions, researchers have attempted to capitalize on the strengths of both Transformers and CNNs. Many studies have investigated combinations of CNNs and Transformers in the context of medical image analysis, which involves not only regional correlations but also subtle textures.

***Transformers with Graphs*:** Using Transformers along with graphs is a common approach in medical image analysis. The basic idea behind graph learning is to obtain a concise representation for each sample, such as embeddings while preserving the inherent relationships between samples through the use of a data graph (Chen et al. 2020). Because of their attention-based nature, Transformers are well-suited for tasks involving graph data, such as node feature aggregation and node relationship computation.

The following results were drawn on the use of Transformers in medical picture classification tasks:

Transformers have shown better performance comparable to convolutional neural network across a wide range of tasks.

- Transformers perform most effectively when applied to huge collection of datasets; however, their implementation, particularly in medical image analysis, is somewhat limited. This limitation can be improved using pre-training techniques.
- Training Transformers on large images imposes a significant computational burden. As a result, increasing efficiency necessitates strategies such as model complexity reduction and the development of lightweight models.
- The growing emphasis on hybrid Transformers is important, given their ability to combine the strengths of conventional networks.

2.5.2 Transformer for Segmentation

Transformer-based methods have been used for various segmentation tasks, such as brain tumor/tissue segmentation, thoracic multi-organ segmentation (Huang et al. 2021), cardiac segmentation (Gao et al. 2023), and abdominal multi-organ segmentation (Yao et al. 2022). Table 2.2 systematically shows details of segmentation approaches.

2.5.2.1 Hybrid Transformers

The current development of combining Transformers with the standard U-shaped model emphasizes three key aspects:

i. Comprising Transformer layers at various stages for the U-shaped design.
ii. Employing various approaches for integrating Transformers and CNNs.
iii. Utilizing attention mechanisms or multi-scale attributes.

TABLE 2.2
Applications of Transformers in Segmentation of Medical Images

Reference	Models	Types	Medical Applications
(Yu et al. 2023)	U-NesT	CT, MRI	Abdominal multiorgan and kidney along with whole brain segmentation
(Sanderson and Matuszewski 2022)	FCN-Transformer	Colonoscopy	CRC segmentation
(Li et al. 2021)	GT U-Net	X-ray	Tooth root canal segmentation
(Chen et al. 2021a)	TransUnet	MRI, CT	MRI and CT cardiac segmentation
(Dan et al. 2023)	Swin Unet	CT scan	Segmentation of abdominal multi-organ

2.5.2.2 Transformer Location in U-Shaped Architecture

One approach involves inserting a complete Transformer between the encoder and decoder sections to establish long-term dependencies in advanced vision ideas. For instance, Chen et al. (2021a) introduced TransUNet, which combines a CNN for high-resolution spatial features with a Transformer for global context encoding.

2.5.3 Image Synthesis

Image synthesis remains extremely challenging in the medical field considering inter-subject variations and the potential impact of morphological anatomical hallucinations on diagnostic tasks. Generative adversarial learning emerged as extensively utilized to address image generation challenges, leading to the integration of Transformers into a generative adversarial training framework. Two methods were introduced for medical image reconstruction: a 3D Transformer GAN for PET images (Luo et al. 2021) and SLATER, an unsupervised technique for MRI synthesis (Korkmaz et al. 2022). Supervised reconstruction techniques in MRI have traditionally been trained on paired undersampled and fully sampled data with additional supervision from the imaging operator. However, we have developed a new approach called Zero-Shot Learned Adversarial TransformERs that reduces the requirement for supervision. It uses a deep adversarial network and cross-attention Transformers to map noise and latent variables to coil-combined MR images. The deep image prior framework, which combines untrained MRI priors with the imaging operator during inference, is not ideal for capturing long-range interactions, and priors derived from randomly initialized networks may produce unsatisfactory results. Therefore, SLATER is a superior unsupervised MRI reconstruction approach that addresses these issues.

2.5.4 Detection

The definition and application of the term "detection" vary in the technical and clinical domains. It usually refers to inspecting for the presence of diseases or abnormalities in technical contexts, whereas it frequently refers to diagnosis or disease classification in clinical settings. In the field of computer vision, detection is a crucial process that involves accurately identifying objects in images and precisely classifying them with high confidence.

TABLE 2.3
Applications of Transformers in Detection of Medical Images

Reference	Method	Category	Application
(Shang et al. 2022)	IHD	CT	Brain injury hemorrhage detection
(Mathai et al. 2022)	DETR	MRI	Lymphoproliferative diseases detection
(Niu and Wang 2022)	UCLT	CT	Lung nodule detection
(Shen et al. 2021)	COTR	Colonoscopy	CRC detection

Medical image recognition absolutely requires the use of Transformers in combination with CNN modules. A CNN extracts properties from biological images, and the Transformer architecture vastly improves the collected parameters for subsequent identification. Kong et al. (2021) introduced CT-CAD, a context-aware hybrid Transformer model that has proven highly effective in identifying chest abnormalities in X-ray images. As shown in Table 2.3, the Transformer used for medical image detection is an essential component for accurate and reliable identification.

2.6 DISCUSSION

The discussion section addresses the effective Transformers implementation in various aspects of biomedical image analysis in a variety of fields. Despite such achievements, the application of deep learning methods and models in clinical settings faces challenges, with label scarcity being a major concern, especially in tasks requiring pixel-level precision labeling. Learning from noisy labels and designing versatile computer-aided diagnosis methods that employ multi-modality clinical data in a multi-task manner are also challenges.

2.7 DIFFERENT LEARNING SCENARIOS FOR TRANSFORMERS

Multi-task Learning: Multi-task training, which involves joint segmentation and classification, improves model generalizability. Multi-Task TransUNet models excel at skin lesion segmentation and classification, achieving cutting-edge performance and efficiency gains.

2.7.1 Multi-Modal Learning

Collecting data from multiple modalities improves diagnostic abilities. Transformer-based models capture relationships between sequences in multi-modal images, demonstrating enhanced efficacy and performance.

2.7.2 Weakly Supervised Learning

Transformers, in weakly supervised conditions, demonstrate effectiveness. Transformers, for example, use multiple instance learning to create correlations between instances, resulting in stronger high-level representations and bag-level predictions.

2.7.3 Self-Supervised Learning

Self-supervised learning is beneficial in overcoming the challenge of limited annotated data. Pre-training Transformer models using self-supervised tasks enable improved downstream task performance, outperforming traditional CNNs.

2.8 CONCLUSIONS

The examination of papers in this chapter clearly demonstrates the widespread integration of Vision Transformers across various domains in medical imaging. In summary, this provides a comprehensive review of Transformer utilization in biomedical images. We discuss the fundamental views of Transformer models performance before providing an extensive literature evaluation on biomedical image applications. We focus on Transformer applications in biomedical image segmentation, recognition, classification, restoration, synthesis, and registration. We provide a summary of each application, identify application-specific challenges, offer solutions, and outline recent developments. In spite of their outstanding performance, we believe Transformers have a lot of undiscovered potential in biomedical imaging. We expect that this chapter will serve as a guide for readers as they work to advance this field.

REFERENCES

Bahdanau, Dzmitry, Kyunghyun Cho, and Yoshua Bengio. 2014. "Neural Machine Translation by Jointly Learning to Align and Translate." https://doi.org/10.48550/ARXIV.1409.0473

Brown, Tom B., Benjamin Mann, Nick Ryder, Melanie Subbiah, Jared Kaplan, Prafulla Dhariwal, Arvind Neelakantan, et al. 2020. "Language Models Are Few-Shot Learners." arXiv. https://doi.org/10.48550/arXiv.2005.14165

Chen, Fenxiao, Yuncheng Wang, Bin Wang, and C.-C. Jay Kuo. 2020. "Graph Representation Learning: A Survey." *APSIPA Transactions on Signal and Information Processing* 9 (1). https://doi.org/10.1017/ATSIP.2020.13

Chen, Jieneng, Yongyi Lu, Qihang Yu, Xiangde Luo, Ehsan Adeli, Yan Wang, Le Lu, Alan L. Yuille, and Yuyin Zhou. 2021a. "TransUNet: Transformers Make Strong Encoders for Medical Image Segmentation." arXiv. https://doi.org/10.48550/arXiv.2102.04306

Chen, Junyu, Yufan He, Eric C. Frey, Ye Li, and Yong Du. 2021b. "ViT-V-Net: Vision Transformer for Unsupervised Volumetric Medical Image Registration." arXiv. http://arxiv.org/abs/2104.06468

Dai, Yin, Yifan Gao, and Fayu Liu. 2021. "TransMed: Transformers Advance Multi-Modal Medical Image Classification." *Diagnostics* 11 (8): 1384. https://doi.org/10.3390/diagnostics11081384

Dan, Yongping, Weishou Jin, Zhida Wang, and Changhao Sun. 2023. "Optimization of U-Shaped Pure Transformer Medical Image Segmentation Network." *PeerJ Computer Science* 9 (August): e1515. https://doi.org/10.7717/peerj-cs.1515

Dosovitskiy, Alexey, Lucas Beyer, Alexander Kolesnikov, Dirk Weissenborn, Xiaohua Zhai, Thomas Unterthiner, Mostafa Dehghani, et al. 2021. "An Image Is Worth 16x16 Words: Transformers for Image Recognition at Scale." arXiv. https://doi.org/10.48550/arXiv.2010.11929

Gao, Yunhe, Mu Zhou, Di Liu, Zhennan Yan, Shaoting Zhang, and Dimitris N. Metaxas. 2023. "A Data-Scalable Transformer for Medical Image Segmentation: Architecture, Model Efficiency, and Benchmark." arXiv. https://doi.org/10.48550/arXiv.2203.00131

Hu, Jie, Li Shen, Samuel Albanie, Gang Sun, and Enhua Wu. 2020. "Squeeze-and-Excitation Networks." *IEEE Transactions on Pattern Analysis and Machine Intelligence* 42 (8): 2011–23. https://doi.org/10.1109/TPAMI.2019.2913372

Huang, Xiaohong, Zhifang Deng, Dandan Li, and Xueguang Yuan. 2021. "MISSFormer: An Effective Medical Image Segmentation Transformer." arXiv. https://doi.org/10.48550/arXiv.2109.07162

Kong, Qiran, Yirui Wu, Chi Yuan, and Yongli Wang. 2021. "CT-CAD: Context-Aware Transformers for End-to-End Chest Abnormality Detection on X-Rays." In *2021 IEEE International Conference on Bioinformatics and Biomedicine (BIBM)*, 1385–88. Houston, TX, USA: IEEE. https://doi.org/10.1109/BIBM52615.2021.9669743

Korkmaz, Yilmaz, Salman UH Dar, Mahmut Yurt, Muzaffer Özbey, and Tolga Çukur. 2022. "Unsupervised MRI Reconstruction via Zero-Shot Learned Adversarial Transformers." arXiv. https://doi.org/10.48550/arXiv.2105.08059

Krizhevsky, Alex, Ilya Sutskever, and Geoffrey E Hinton. 2012. "ImageNet Classification with Deep Convolutional Neural Networks." In *Advances in Neural Information Processing Systems*. Vol. 25. Curran Associates, Inc. https://proceedings.neurips.cc/paper_files/paper/2012/hash/c399862d3b9d6b76c8436e924a68c45b-Abstract.html

Li, Yunxiang, Shuai Wang, Jun Wang, Guodong Zeng, Wenjun Liu, Qianni Zhang, Qun Jin, and Yaqi Wang. 2021. "GT U-Net: A U-Net Like Group Transformer Network for Tooth Root Segmentation." *Machine Learning in Medical Imaging: 12th International Workshop, MLMI 2021, Held in Conjunction with MICCAI 2021*, 12966:386–95. https://doi.org/10.1007/978-3-030-87589-3_40

Liu, Wanli, Chen Li, Md Mamunur Rahaman, Tao Jiang, Hongzan Sun, Xiangchen Wu, Weiming Hu, et al. 2022. "Is the Aspect Ratio of Cells Important in Deep Learning? A Robust Comparison of Deep Learning Methods for Multi-Scale Cytopathology Cell Image Classification: From Convolutional Neural Networks to Visual Transformers." *Computers in Biology and Medicine* 141 (February): 105026. https://doi.org/10.1016/j.compbiomed.2021.105026

Liu, Ze, Yutong Lin, Yue Cao, Han Hu, Yixuan Wei, Zheng Zhang, Stephen Lin, and Baining Guo. 2021. "Swin Transformer: Hierarchical Vision Transformer Using Shifted Windows." arXiv. https://doi.org/10.48550/arXiv.2103.14030

Luo, Yanmei, Yan Wang, Chen Zu, Bo Zhan, Xi Wu, Jiliu Zhou, Dinggang Shen, and Luping Zhou. 2021. "3D Transformer-GAN for High-Quality PET Reconstruction." In *Medical Image Computing and Computer Assisted Intervention – MICCAI 2021*, edited by Marleen De Bruijne, Philippe C. Cattin, Stéphane Cotin, Nicolas Padoy, Stefanie Speidel, Yefeng Zheng, and Caroline Essert, 12906:276–85. Lecture Notes in Computer Science. Cham: Springer International Publishing. https://doi.org/10.1007/978-3-030-87231-1_27

Lyu, Qing, Sanjeev V. Namjoshi, Emory McTyre, Umit Topaloglu, Richard Barcus, Michael D. Chan, Christina K. Cramer, et al. 2022. "A Transformer-Based Deep-Learning Approach for Classifying Brain Metastases into Primary Organ Sites Using Clinical Whole-Brain MRI Images." *Patterns* 3 (11): 100613. https://doi.org/10.1016/j.patter.2022.100613

Mathai, Tejas Sudharshan, Sungwon Lee, Daniel C. Elton, Thomas C. Shen, Yifan Peng, Zhiyong Lu, and Ronald M. Summers. 2022. "Lymph Node Detection in T2 MRI with Transformers." In *Medical Imaging 2022: Computer-Aided Diagnosis*, edited by Khan M. Iftekharuddin, Karen Drukker, Maciej A. Mazurowski, Hongbing Lu, Chisako Muramatsu, and Ravi K. Samala, 62. San Diego, United States: SPIE. https://doi.org/10.1117/12.2613273

Niu, Chuang, and Ge Wang. 2022. "Unsupervised Contrastive Learning Based Transformer for Lung Nodule Detection." *Physics in Medicine & Biology* 67 (20): 204001. https://doi.org/10.1088/1361-6560/ac92ba

Sanderson, Edward, and Bogdan J. Matuszewski. 2022. "FCN-Transformer Feature Fusion for Polyp Segmentation." In *Medical Image Understanding and Analysis*, edited by

Guang Yang, Angelica Aviles-Rivero, Michael Roberts, and Carola-Bibiane Schönlieb, 13413:892–907. Lecture Notes in Computer Science. Cham: Springer International Publishing. https://doi.org/10.1007/978-3-031-12053-4_65

Shamshad, Fahad, Salman Khan, Syed Waqas Zamir, Muhammad Haris Khan, Munawar Hayat, Fahad Shahbaz Khan, and Huazhu Fu. 2023. "Transformers in Medical Imaging: A Survey." *Medical Image Analysis* 88 (August): 102802. https://doi.org/10.1016/j.media.2023.102802

Shang, Fangxin, Siqi Wang, Xiaorong Wang, and Yehui Yang. 2022. "An Effective Transformer-Based Solution for RSNA Intracranial Hemorrhage Detection Competition." arXiv. https://doi.org/10.48550/arXiv.2205.07556

Shen, Zhiqiang, Rongda Fu, Chaonan Lin, and Shaohua Zheng. 2021. "COTR: Convolution in Transformer Network for End to End Polyp Detection." In *2021 7th International Conference on Computer and Communications (ICCC)*, 1757–61. Chengdu, China: IEEE. https://doi.org/10.1109/ICCC54389.2021.9674267

Stegmuller, Thomas, Behzad Bozorgtabar, Antoine Spahr, and Jean-Philippe Thiran. 2023. "ScoreNet: Learning Non-Uniform Attention and Augmentation for Transformer-Based Histopathological Image Classification." In *2023 IEEE/CVF Winter Conference on Applications of Computer Vision (WACV)*, 6159–68. Waikoloa, HI, USA: IEEE. https://doi.org/10.1109/WACV56688.2023.00611

Vaswani, Ashish, Noam Shazeer, Niki Parmar, Jakob Uszkoreit, Llion Jones, Aidan N. Gomez, Lukasz Kaiser, and Illia Polosukhin. 2017. "Attention Is All You Need." https://doi.org/10.48550/ARXIV.1706.03762

Yao, Chang, Menghan Hu, Qingli Li, Guangtao Zhai, and Xiao-Ping Zhang. 2022. "Transclaw U-Net: Claw U-Net With Transformers for Medical Image Segmentation." In *2022 5th International Conference on Information Communication and Signal Processing (ICICSP)*, 280–84. https://doi.org/10.1109/ICICSP55539.2022.10050624

Yu, Xin, Qi Yang, Yinchi Zhou, Leon Y. Cai, Riqiang Gao, Ho Hin Lee, Thomas Li, et al. 2023. "UNesT: Local Spatial Representation Learning with Hierarchical Transformer for Efficient Medical Segmentation." arXiv. https://doi.org/10.48550/arXiv.2209.14378

Zhang, Fan, Tengfei Xue, Weidong Cai, Yogesh Rathi, Carl-Fredrik Westin, and Lauren J. O'Donnell. 2022. "TractoFormer: A Novel Fiber-Level Whole Brain Tractography Analysis Framework Using Spectral Embedding and Vision Transformers." In *Medical Image Computing and Computer Assisted Intervention – MICCAI 2022: 25th International Conference, Singapore, September 18–22, 2022, Proceedings, Part I*, 196–206. Berlin, Heidelberg: Springer-Verlag. https://doi.org/10.1007/978-3-031-16431-6_19

Zhang, Lei, and Yan Wen. 2021. "A Transformer-Based Framework for Automatic COVID19 Diagnosis in Chest CTs." In *2021 IEEE/CVF International Conference on Computer Vision Workshops (ICCVW)*, 513–18. Montreal, BC, Canada: IEEE. https://doi.org/10.1109/ICCVW54120.2021.00063

3 Future of Learning

Adaptive Learning Systems Based on Language Generative Models in Higher Education

Cristina Dumitru
The National University of Science and Technology POLITEHNICA Bucharest, Pitești University Centre, Romania

3.1 INTRODUCTION: EDUCATIONAL BACKGROUND

The integration of technology into education, particularly in higher education, is an ongoing and expanding trend that is anticipated to endure in the future. Artificial Intelligence (AI) in education has experienced rapid growth, significantly impacting teaching, learning, and research experiences. Due to technological advancements, AI is reshaping the higher education learning landscape, and this transformation is being reciprocated by the learners themselves (Alam and Mohanty 2022). Numerous studies have delved into the implication of AI in education, with a specific focus on adaptive learning, smart campuses, tutoring robots, virtual assistants, and generative AI models such as chatbots (Wang et al. 2023). Technology is increasingly recognized as a promising solution to overcome barriers in learning, and the imperative to embrace technological advancements is widely acknowledged across educational institutions (Wang et al. 2023). A literature review conducted by Hannan and Liu (2023) supports the beneficial application of AI in enhancing student learning experiences, providing educational support, and improving the enrollment process.

AI is changing the way education is delivered and has erased the boundaries of time and places of learning and is currently continuing to transform and define the contexts of higher education. Rather than replacing traditional lectures, AI in education aims to enhance the efficiency of students' knowledge acquisition and engagement by providing access to information and tools to navigate the vast amount of available information.

AI provides tools for autonomous learning and facilitates the co-creation of adaptive learning paths. Adaptive learning with AI takes into consideration students' prior learning experiences, continuously and tirelessly monitors the ongoing learning process, and assesses the quality of knowledge assimilation. Each student within the adaptive learning process follows a personalized learning path customized to their

DOI: 10.1201/9781032644509-3

potential, resources, educational needs, objectives, motivations, and individual abilities to comprehend, process, and engage with educational content. Therefore, ALSs leverages AI and data analytics to customize learning experiences to the unique needs and preferences of individual students, ushering in an era of personalized education.

This chapter, based on the input from recent studies, discusses the implications of ALSs in higher education. It aims to explore and evaluate how the solutions proposed by ALSs can play a pivotal role in assisting both teachers and students in the teaching and learning processes. Based on specific prompts, ALSs are closer to provide an adaptive pathway in higher education. The incorporation of ALSs in higher education is a relatively recent development. As a result, numerous educators find themselves grappling with the challenge of identifying efficient ways and effective methods to integrate AI and ALSs into higher education, aiming to enhance learning efficiency and facilitate educational experiences and motivation for their students. This chapter was based on a systematic literature review of existing publications between 2020 and 2023.

3.2 ADAPTIVE LEARNING SYSTEMS IN HIGHER EDUCATION

ALSs encompass any educational system designed to consider the unique characteristics of individual learners, including their potential as well as their barriers in learning. As emphasized by Oxman, Wong, and Innovations (2014), the fundamental goal of adaptive learning is to enhance students' educational outcomes. These systems aim to adapt the content and processes of education to the learner's knowledge and goals. The efficiency of ALSs lies in their ability to distinguish what students already know from what they don't know, utilizing individual student characteristics to suggest suitable learning materials.

Advancements in information technology, coupled with increased interest, enable the cost-effective implementation of intricate tasks in AI and large-scale data processing. This progress has become particularly relevant with the widespread adoption of e-learning, creating a new software product niche, specifically in the domain of adaptive learning services and systems.

In the context of the increased accessibility of higher education, universities have sought approaches to cater to each student, fostering effective learning (Neves and Hillman 2017). Many companies producing digital products today incorporate elements of adaptive learning into their systems. When developing an adaptive educational system, three key questions are initially addressed: what is modeled (which variables are included in the adaptation model), how it is modeled (based on what interactions between variables the student receives educational material), and how the operation of the adaptation model is supported (Johanes and Lagerstrom 2017). Subsequently, one of the three scenarios is implemented, where the object of adaptation can be: (1) content, (2) tasks, or (3) the order of presentation of educational materials. According to Pearson's research, the target for adaptation in educational systems varies depending on the level of education (Hogan and Sellar 2021). Thus, within the framework of general school education, adaptive systems are more frequently used based on changing content, while for the universities, educational

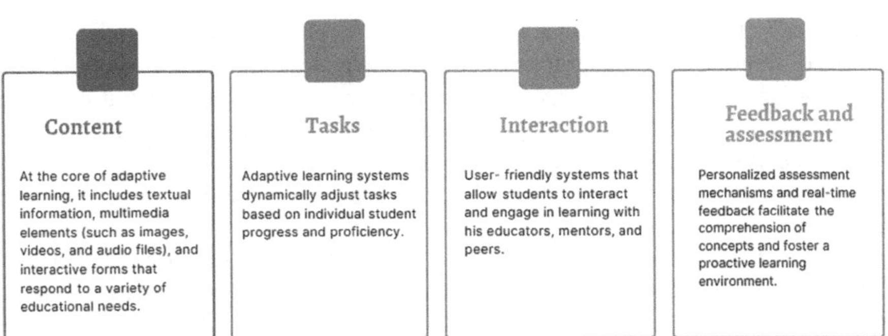

FIGURE 3.1 Features of adaptive learning systems

Adapted from Johanes and Lagerstrom 2017.

systems that adjust the order of presentation of educational materials are more popular. Figure 3.1 illustrates the key adaptation elements for personalized learning experiences in the context of higher education.

The degree and methods of adaptation depicted in Figure 3.1 vary, depending on the specific context of software solution application and features of the educational process in which they are incorporated. In higher education, the active utilization of adaptive technologies remains limited, as their implementation implies, to a large extent, educational reform. The adoption of adaptive learning technologies in higher education is constrained primarily because their implementation necessitates a substantial shift in educational paradigms.

The anticipated effects resulting from the introduction of ALSs are impressive, including increased motivation (Kochurina, Fedorova, and Zakharov 2022; Orji and Vassileva 2021; Johanes and Lagerstrom 2017) and engagement of students in learning (Liu et al. 2022), improved educational results (Wang et al. 2020), better preparation of students for the labor market (Wang et al. 2020), and a reduction in the dropout rate (Owens 2021).

3.3 FOUNDATIONS OF ADAPTIVE LEARNING: THE PEDAGOGICAL APPROACH, THE ROLE OF ARTIFICIAL INTELLIGENCE, DATA ANALYTICS, AND LEARNING ANALYTICS

ALSs have gained widespread adoption in Western universities, with institutions turning to the services of companies such as Knewton[1] and Cerego.[2] Companies like Cerego offer products that facilitate the implementation of adaptive learning across various academic disciplines, complementing traditional teaching methods. However, their expansion to other universities is challenged by the existing electronic educational infrastructure of respective institutions (Кречетов and Романенко 2020).

The ALS in which the educational content is adapted operates on the following principles. Firstly, the system records and accurately assesses the student's response to a given task. In the event of an error, the ALS is able to respond dynamically by providing hints, instant feedback, or supplementary educational materials to guide the student toward a better understanding of the concept. These additional resources can take various forms, such as instructive video lectures, comprehensive texts, and step-by-step instructions. Furthermore, students may have the opportunity to seek clarification or engage in further discussions with an educator, a tutor, or a peer, ensuring a personalized and interactive learning experience.

This approach ensures that the ALS not only identifies areas of difficulty but also actively supports students in overcoming challenges. The goal is to create a responsive and tailored learning environment that optimally addresses the individual needs and learning pace of each student. Through this adaptive content model, students can benefit from a comprehensive and personalized educational experience, enhancing their comprehension and mastery of the subject matter.

Вилкова and Лебедев (2020) provide an example of educational content adaptation using the platform Smart Sparrow.[3] This platform offers instant feedback for any interaction with educational materials. On this platform, three levels of adaptation are possible: feedback, curriculum modification, and the educator's ability to adjust the knowledge transfer process. Another example is the Plario adaptive mathematics learning system, which assesses student progress using the Bayesian Knowledge Tracing (BKT) method, tailoring the learning path based on the student's preparedness level (Вилкова and Лебедев 2020).

Other ALSs operate on a different principle, analyzing student responses and prompting them to complete the next task upon providing an answer, instead of directing them to additional materials (Edelen and Reeve 2007). These systems use Item Response Theory (IRT) (Lord 2012). Within the IRT framework, the probability of correctly completing a task is determined based on the student's preparedness level and task parameters (van der Linden and Hambleton 2013). For an efficient utilization, this ALS creates a task bank with known and fixed difficulty levels, allowing targeted adaptation in subsequent tasks if a student answers a question correctly. If a student answers a question correctly, it indicates their level of mastery, allowing for targeted adaptation in subsequent tasks.

ALSs utilize algorithms and predictive analytics to continuously collect data on the learning process and adjust the delivery of educational materials accordingly. The process involves three main steps: data collection, data analysis, and the provision of educational materials to the learner. Variables such as student responses to assignments, the number of attempts, the frequency of errors, interactive resource usage, and reading materials contribute to the collected data. The system analyzes this information to discern the student's existing knowledge and knowledge gaps, tailoring content to the learner's needs.

Algorithmic adaptivity is a form of adaptability that relies on one or more algorithms to address fundamental questions: firstly, assessing what the student already knows, and secondly, determining what the next learning experience should be. This method enables algorithms to choose the most appropriate learning experience at the right moment for each student. ALSs provide several tools for enhancing specific educational experience, see Table 3.1.

TABLE 3.1
Adaptive Learning Technologies Analysis

Tool Type	Description
Communication and cooperation Functions	– Student-teacher communication – Collaborative tools for students
Content model and customization Features	– Monitoring specific student answers – Presentation of unique tips, feedback, links for resources on a specific topic
Benchmark assessment	– Change of items depending on how individual students answer each question
Trainer (Practice Engine)	– Use of adaptive estimation with a pool of questions of varying difficulty levels matching the content just learned – Questions are presented post-class, and students demonstrate mastery by answering questions – Continues until the student sufficiently answers difficult questions and then moves on to the next skill
Adaptive structuring	– Works with Big Data, continuously collecting and analyzing electronic traces – Automatically adjusts what the learner sees, the order of skill representations, and the type of content received – Utilizes algorithms and predictive analytics for continuous data collection, changing the student's next steps based on the analysis

Based on Добрица and Горюшкин 2019

One widely used algorithm in this context is Knewton platform, an example of ALS that provides support in adapting the educational process to individual student characteristics, by assessing students' strengths, identifying areas requiring additional attention, and recommending modifications to enhance the likelihood of future success throughout the course (Johnson and Samora 2016). Consequently, universities and educational applications collaborating with Knewton get insights into student learning processes. Progression to the next topic is contingent upon mastering the preceding one, ensuring personalized and effective learning experiences (Bernacki and Walkington 2018).

These tools cater to different aspects of adaptive learning, providing diverse approaches to enhance the learning experience for students.

3.4 ROLE OF TECHNOLOGY IN SHAPING EDUCATION

"Despite risks, some educators see huge potential in using artificial-intelligence chatbots to enhance teaching and learning" (Extance 2023, 474). Researchers, educators, and organizations are exploring methods to transform expansive language models into reliable and precise "thought partners" for educational purposes (Extance 2023). The emphasis lies in fostering and motivating personalized learning within the higher education setting. This involves the utilization of advanced language models to enhance the educational experience through tailored and individualized learning approaches.

TABLE 3.2
Characteristics of Adaptive Learning Technology

Characteristics	Description
Objectives of technology use in education	– Providing support in studying, independent work, self-control, research, and project activities – Developing and improving the ability to work independently and acquire knowledge – Maximizing adaptation of the educational process to individual student characteristics
Technology features	– Students plan their own course of learning – Teacher uses flexible settings of pace and mode of educational work – Monitoring of educational work – Special methods created for students' independent work – Central focus on student's individual characteristics for qualitative change – Regular interaction for studying, applying, and consolidating educational material with mutual control

In response to the ability of AI chatbots to generate articulate and well-researched essays in response to assignment questions, educators had to rethink the assessment approach.[4] Simultaneously, some educational institutions are embracing the "flipped" classroom model, wherein students' complete assignments during school hours after independently studying a subject as a home assignment. This transformation accompanies the evolving landscape of education, driven by advancements in AI technology. In line with these developments, the structure of university lectures and of the learning process are being redefined to accommodate the growing influence of AI and ALSs in education.

For practitioners, embracing ALS necessitates a thoughtful integration into existing educational frameworks. Educators need to strike a balance between leveraging the benefits of adaptive learning and maintaining a pedagogical approach that values human interaction, critical thinking, and creativity.

Table 3.2 provides an overview of the key characteristics of adaptive learning technology, highlighting its various features and functionalities.

Adaptive learning technology aims to enhance educational effectiveness by understanding individual student needs, prior knowledge, and learning abilities. Technology can be used in education to improve educational experience, providing differentiated tasks, enabling independent work, and facilitating effective planning and organization of educational activities. Adaptive learning technology, through the use of technology, becomes a flexible system that accommodates diverse teaching methods and allows for variations in the duration and sequence of learning steps.

Some proposed uses of technology in higher education, as suggested by Назарова (2015) are as follows:

- *Managing the process of acquiring educational content*, by employing strategic approaches to ensure comprehensive understanding and retention of academic content Matayoshi et al. (2019). This includes employing technological tools that develop efficient study habits, utilize appropriate resources, and implement tailored learning techniques.
- *Providing opportunities for practical application of acquired knowledge*, important for bridging the gap between theory and real-world scenarios. This involves actively using learned concepts and skills in practical situations, enhancing the ability to solve problems, make informed decisions, and contribute effectively in professional environments.
- *Searching and processing the new information*, which involves utilizing various research methods, information sources, and digital tools to gather, evaluate, and incorporate the latest insights into one's knowledge base.
- *Organizing study groups (peer learning, mentee, or tutoring activities)*, by creating an environment conducive to collaboration, fostering open communication, and establishing clear roles and responsibilities for peer learning, mentoring, or tutoring activities.
- *Organizing independent learning activities*, which empowers students to take initiative in their learning journey, pursue personal development goals, and enhance overall autonomy. Various technical tools that allow the practice of self-discipline, time management, and goal-setting are available.
- *Organizing educational and cognitive activities*, which involves planning and structuring learning experiences to optimize comprehension and memorization. This includes selecting appropriate instructional methods, resources, and assessment strategies tailored to individual and collective needs.
- *Conducting research activities*, engaging in research activities involves systematic inquiry, data collection, and critical analysis. This skill enhances the capacity for independent investigation, the development of evidence-based perspectives, and the contribution to academic and professional knowledge. Currently, AI provides a great number of tools which facilitates the research activities.

Despite its potential, its incorporation into higher education has been limited due to technical complexities and a lack of readiness among educational institutions. Overall, the implementation of adaptive learning technology is portrayed as a dynamic and responsive approach to education, promoting individualized learning experiences and student engagement. In the context of innovative learning systems, each educational technology has a distinct role and significance.

3.5 CHALLENGES OF IMPLEMENTING ADAPTIVE LEARNING IN HIGHER EDUCATION

In the context of the massification of higher education, universities need to find an approach to each of the students, helping them to study effectively (Neves and Nick Hillman 2017). Several voices refer to adaptive learning as one of the "big challenges" facing the educational system (Johanes and Lagerstrom 2017).

However, despite its current popularity, for a long-time ALSs failed to have significant distribution and implementation in education. Several factors contributed to this delayed adoption, including (i) the insufficient level of computer performance of teachers and students, hindering the implementation of new technology, (ii) the complexity of software implementation, (iii) the insufficient development and prevalence of e-learning technologies, and (iv) limited availability of data-driven evidence showcasing the efficacy of ALSs in higher education.

The aim of the adaptive learning is to build an optimal trajectory for a student to pursue a study module. A module is a logically complete minimal unit of an educational content that reveals one or more terms or concepts and is in connection with other units, based on a constructivist approach to learning (Dewey 1910). In higher education, study programs are organized considering obtaining the maximum level of knowledge at the time of completion of the course with a minimum time for studying course modules (Кречетов and Романенко 2020). ALSs support educators in organizing educational content in modules, containing not only text but also graphics, video, or audio files, as well as any other interactive forms of presentation of information.

At the same time, the COVID-19 pandemic has significantly changed the higher education system. At the beginning of 2020, students around the world began studying online, traditional lectures and seminars transitioned to a new format. In a short period of time, online technologies became widespread, but the question of their quality and effectiveness remains open (García-Morales, Garrido-Moreno, and Martín-Rojas 2021).

Although the idea of adaptive learning seems promising, there is limited evidence of its effectiveness. The lack of consensus on the effectiveness of ALSs in higher education is attributed not only to empirical data limitations but also to inherent challenges in adaptive learning. These limitations include a restricted range of disciplines for adaptation, complexities in measuring intricate constructs, the necessity for substantial financial and time investments, and limited adaptation capabilities (Вилкова and Лебедев 2020). Currently, ALSs focus on medical and STEM (Science, Technology, Engineering, Math) disciplines. In contrast, in social sciences and humanities, there is a limited application of AL, with examples such as the CogBooks[5] system for history courses and the MyLab[6] platform offering materials in social sciences. The challenge lies in the difficulty of translating a particular course into an adapted format, resulting in uncertainty regarding the effectiveness of adaptive learning across disciplines.

Another challenge associated with ALSs is their high costs, both in terms of time and in terms of financial and human resources (Li et al. 2022), due to the fact that it requires the development of a substantial amount of educational material that will be suitable for students' specific needs. Implementing such a capability could facilitate accurate assessments of students' capabilities and preferences, eliminating the need to create a new student model for each new educational course. However, the lack of universally accepted standards for the development of adaptive systems that consider their unique characteristics poses a significant challenge (Вилкова and Лебедев 2020).

Perrotta and Williamson (2016) note that in some cases, the mass introduction of adaptive learning through AI may not reduce educational inequality, but, on the contrary, may increase the gap between students with different socio-economic status. Additionally, large companies creating ALSs are often driven more by profit motives than pedagogical principles (Вилкова and Лебедев 2020).

Overcoming these barriers is crucial for realizing the full potential of ALSs in education. Addressing issues related to technology infrastructure, software complexity, and the need for robust data evidence will pave the way for a more widespread and effective integration of adaptive learning strategies in educational settings. As these challenges are progressively addressed and technological advancements continue, ALSs is expected to play an increasingly significant role in reshaping and enhancing the educational experience for both educators and students alike. However, Popenici (2022) expresses his doubt regarding the capabilities of technology to address all major challenges we are currently facing.

The integration of ALSs based on AI in higher education must be grounded on data. Few studies show the advantage of AI adaptive educational systems, and existing work does not allow clear conclusions about the effectiveness of adaptive learning.

3.6 CONCLUSION: IMPLICATIONS FOR PRACTICE, ETHICAL CONSIDERATIONS IN ADAPTIVE LEARNING, AND FUTURE DIRECTIONS FOR RESEARCH

This chapter presents the results of a literature review that aimed to investigate the ways of using ALSs based on AI as a solution to enhance students' learning during higher education. The concept of personalized education and the need to adapt the educational process to individual characteristics of the student have been longstanding considerations among educators. With the advancement of computer technology, interest in adaptive learning expanded as a promising solution in higher education. As digital technologies become crucial tools for providing access to educational resources, especially in the context of the massification of higher education, ALS have emerged. However, the transformative impact of adaptive learning on student effectiveness remains uncertain. Some researchers even suggest that the use of adaptive technologies might increase educational inequality. Dwivedi et al. (2023) urge the need to address the ethical, legal, and social challenges posed by the misuse and abuse of generative AI, advocating for considering its potential harms.

This review delves into the processes of adaptive learning algorithms, ongoing research to evaluate their effectiveness, and the challenges they face. The primary conclusion drawn is neither decisively positive nor negative, as existing empirical studies lack a consensus on the effectiveness of adaptive learning. The central question revolves around whether students can enhance their educational outcomes when the educational process adapts and accommodates their individual characteristics.

Despite its challenges and inconsistent mechanism of its implementation in higher education, ALSs are expanding and refining making it possible to make more accurate predictions about student learning behaviors.

In conclusion, the use of ALSs holds significant implications for educational practice, posing both opportunities and challenges. The personalized and adaptive nature of these systems allows for tailored learning experiences, potentially enhancing student engagement and outcomes. However, ethical considerations must be carefully addressed, particularly concerning data privacy, algorithmic biases, and the potential for widening educational inequalities. As technology continues to evolve, ethical guidelines should be continuously revisited and updated to ensure responsible and equitable implementation. Looking ahead, future research in the field should focus on refining and validating the effectiveness of adaptive learning algorithms. Additionally, exploring innovative ways to mitigate ethical concerns and addressing the digital divide will be crucial for the continued advancement and responsible adoption of ALS.

NOTES

1 https://support.knewton.com/s/.
2 https://www.cerego.com/.
3 https://www.smartsparrow.com/.
4 "Why teachers should explore ChatGPT's potential – despite the risks." *Nature* vol. 623, 7987 (2023): 457–458.
5 https://www.cogbooks.com/.
6 https://mlm.pearson.com/global/.

REFERENCES

Alam, Ashraf, and Atasi Mohanty. "Foundation for the future of higher education or 'misplaced optimism'? Being human in the age of artificial intelligence." In *Innovations in Intelligent Computing and Communication, ICIICC 2022. Communications in Computer and Information Science*, vol. 1737, edited by M. Panda et al., 17–29. Cham: Springer, 2022. https://doi.org/10.1007/978-3-031-23233-6_2

Bernacki, Matthew L., and Candace Walkington. "The role of situational interest in personalized learning." *Journal of Educational Psychology* 110, no. 6 (2018): 864–881 https://doi.org/10.1037/edu0000250

Dewey, John. *The child and the curriculum*. Chicago: University of Chicago Press, 1910.

Dwivedi, Yogesh K., Nir Kshetri, Laurie Hughes, Emma Louise Slade, Anand Jeyaraj, Arpan Kumar Kar, Abdullah M. Baabdullah et al. ""So what if ChatGPT wrote it?" Multidisciplinary perspectives on opportunities, challenges and implications of generative conversational AI for research, practice and policy." *International Journal of Information Management* 71 (2023): 102642. https://doi.org/10.1016/j.ijinfomgt.2023.102642

Edelen, Maria Orlando, and Bryce B. Reeve. "Applying item response theory (IRT) modeling to questionnaire development, evaluation, and refinement." *Quality of life research* 16, no.1 (March 21, 2007): 5–18. https://doi.org/10.1007/s11136-007-9198-0

Extance, Andy. "ChatGPT has entered the classroom: How LLMs could transform education. " *Nature* 623, no. 7987 (November 15, 2023): 474–477. https://doi.org/10.1038/d41586-023-03507-3

García-Morales, Víctor J., Aurora Garrido-Moreno, and Rodrigo Martín-Rojas. "The transformation of higher education after the COVID disruption: Emerging challenges in an online learning scenario." *Frontiers in Psychology* 12 (February 11, 2021): 616059. https://doi.org/10.3389/fpsyg.2021.616059

Добрица, В. П., and Е. И. Горюшкин. "Применение интеллектуальной адаптивной платформы в образовании." ["Application of an intelligent adaptive platform in education"] *Auditorium* 1, no. 21 (2019): 86–92.

Hannan, Erin, and Shuguang Liu. "AI: New source of competitiveness in higher education." *Competitiveness Review: An International Business Journal* 33, no. 2 (February 7, 2023): 265–279. https://doi.org/10.1108/CR-03-2021-0045

Hogan, Anna, and Sam Sellar. "Pearson's digital transformation and the disruption of public education." In *Digital disruption in teaching and testing*, pp. 107–123. Routledge, 2021.

Johanes, Petr, and Larry Lagerstrom. "Adaptive learning: The premise, promise, and pitfalls." In *2017 ASEE Annual Conference & Exposition* (June 24, 2017). https://doi.org/10.10.18260/1-2--27538

Johnson, David, and Dina Samora. "The potential transformation of higher education through computer-based adaptive learning systems." *Global Education Journal* 2016, no. 1 (2016): 1–17.

Kochurina, Svetlana A., Olga O. Fedorova, and Konstantin P. Zakharov. "Learning Motivation of Students of Adaptive Learning Platforms." In *Proceedings of the Conference "Integrating Engineering Education and Humanities for Global Intercultural Perspectives"*, 499 (July 26, 2022): 303–310. Cham: Springer International Publishing, 2022. https://doi.org/10.1007/978-3-031-11435-9_33

Кречетов, Иван Анатольевич, and Владимир Васильевич Романенко. "Реализация методов адаптивного обучения" ["Implementation of adaptive learning methods"]. Вопросы образования *[Educational Studies]* 2 (2020): 252–277. https://doi.org/10.17323/1814-9545-2020-2-252-277

Li, Xiao, Hanchen Xu, Jinming Zhang, and Hua-hua Chang. "Deep reinforcement learning for adaptive learning systems." *Journal of Educational and Behavioral Statistics* 48, no. 2 (November 3, 2022): 220–243. https://doi.org/10.3102/10769986221129847

Liu, Hsin-Lan, Tao-Hua Wang, Hao-Chiang Koong Lin, Chin-Feng Lai, and Yueh-Min Huang. "The influence of affective feedback adaptive learning system on learning engagement and self-directed learning." *Frontiers in psychology* 13 (April 27, 2022): 858411. https://doi.org/10.3389/fpsyg.2022.858411

Lord, Frederic M. *Applications of item response theory to practical testing problems*. Routledge, 2012.

Matayoshi, Jeffrey, Hasan Uzun, and Eric Cosyn. "Deep (un) learning: Using neural networks to model retention and forgetting in an adaptive learning system." In *Artificial Intelligence in Education: 20th International Conference, AIED*, 1625 (June 21, 2019): 258–269. Springer International Publishing. https://doi.org/10.1007/978-3-030-23204-7_22

Назарова, Г. Н. "Роль образовательных технологий в формировании системы инновационного обучения." ["The role of educational technologies in the formation of a system of innovative education"] Территория науки 1 (2015): 33–37.

Neves, Jonathan, and Nick Hillman. "Student academic experience survey." *Higher Education Policy Institute and Higher Education Academy* 12 (2017): 47–48. https://doi.org/10.1093/ietcom/e90-b.4.885

Orji, Fidelia A., and Julita Vassileva. "Modelling and quantifying learner motivation for adaptive systems: Current insight and future perspectives." In *International Conference on Human-Computer Interaction*, 12793 (July 3, 2021): 79–92. Cham: Springer International Publishing. https://doi.org/10.1007/978-3-030-77873-6_6

Owens, Kevin P. "Competency-based experiential-expertise and future adaptive learning systems." In *International Conference on Human-Computer Interaction*, 12793 (July 3, 2021): 93–109. Cham: Springer International Publishing. https://doi.org/10.1007/978-3-030-77873-6_7

Oxman, Steven, William Wong, and D. Innovations. "White paper: Adaptive learning systems." *Integrated Education Solutions* (2014): 6–7. http://kenanaonline.com/files/0100/100321/DVx_Adaptive_Learning_White_Paper.pdf

Perrotta, Carlo, and Ben Williamson. "The social life of learning analytics: Cluster analysis and the 'performance' of algorithmic education." *Learning, Media and Technology* 43, no. 1 (May 17, 2016): 3–16. https://doi.org/10.1080/17439884.2016.1182927

Popenici, Stefan. *Artificial intelligence and learning futures: Critical narratives of technology and imagination in higher education*. Taylor & Francis, 2022. https://doi.org/10.4324/9781003266563

van der Linden, Wim J., and Ronald K. Hambleton, eds. *Handbook of modern item response theory*. New York: Springer Science & Business Media, 2013. https://doi.org/10.1007/978-1-4757-2691-6

Вилкова, К. А., and Д. В. Лебедев Адаптивное обучение в высшем образовании: за и против. *[Adaptive learning in higher education: pros and cons]*. Москва: НИУ ВШЭ, 2020.

Wang, Shuai, Claire Christensen, Wei Cui, Richard Tong, Louise Yarnall, Linda Shear, and Mingyu Feng. "When adaptive learning is effective learning: Comparison of an adaptive learning system to teacher-led instruction." *Interactive Learning Environments* 31, no. 2 (August 31, 2020): 793–803. https://doi.org/10.1080/10494820.2020.1808794

Wang, Ting, Brady D. Lund, Agostino Marengo, Alessandro Pagano, Nishith Reddy Mannuru, Zoë A. Teel, and Jenny Pange. "Exploring the potential impact of artificial intelligence (AI) on international students in higher education: Generative AI, chatbots, analytics, and international student success." *Applied Sciences* 13, no. 11 (May 31, 2023): 6716. https://doi.org/10.3390/app13116716

4 Artificial Intelligence (AI) Integration in Higher Education

Navigating Opportunities and Ethical Frontiers in Education with Advanced Technologies

Seema Yadav
Bhopal School of Social Sciences, Bhopal, India

4.1 INTRODUCTION

The higher education system has undergone substantial changes as it struggles to provide a viable educational system and transversal capabilities for market demands (Bucea-Manea-Ţoniş et al. 2022). Technology advancements, particularly the expanding impact of artificial intelligence (AI), have brought about an unparalleled upheaval in higher education. Education is only one area in which artificial intelligence has become prevalent. Higher education is being greatly impacted by this technology, which is transforming the way we teach and learn (Solís et al. 2023). Big Data management, student retention, job market readiness, and equipping students with the skills needed for the fourth industrial revolution have all been facilitated by AI in higher education (Slimi and Carballido 2023). In recent years, artificial intelligence in higher education has drawn a lot of interest (Wang and Zhan 2021). As a strategic tool, artificial intelligence systems improve people's lives and society overall. They also mark a new chapter in the evolution of digital technologies and modern civilization as a whole. As a strategic tool, artificial intelligence systems improve people's lives and society overall. They also mark a new chapter in the evolution of digital technologies and modern civilization as a whole (Drach et al. 2023). The COVID-19 epidemic has defined the past year, during which there has been a surge in the use of important emerging technologies in education. To improve and actively influence the learning process in the context of online higher education, technology and education increased their collaboration under these difficult circumstances. Artificial intelligence and extended reality (XR) were available toolkits that might be

DOI: 10.1201/9781032644509-4

used to enhance a student-centered and dynamic teaching approach from an educational standpoint (Rangel-de Lázaro and Duart 2023).

Education institutions need to develop new digital competencies in IA, ML, IoT, 5G, cloud, Big Data, blockchain, data analysis, using MS Office and other apps, MOOCs, simulation software, VR/AR, and gamification (Bucea-Manea-Ţoniş et al., 2022). Applications of AI in higher education for instruction and learning are anticipated to increase dramatically (Wang and Zhan 2021). Artificial intelligence has had a significant influence on higher education and is without a doubt one of the most innovative technologies to support university instruction as well as overall quality (Wang and Zhan 2021). The abundance of information made available to us by the internet and Big Data technologies has had a profound impact on the economy and society. As a result, its use in college and university teaching administration and education has garnered a lot of interest. Artificial intelligence introduces a fresh perspective for the invention of teaching management techniques in colleges and universities with its more sophisticated, individualized, precise, and varied features (Ge and Hu 2020).

Artificial intelligence is the capacity of machines and computer systems to simulate certain cognitive processes of humans, including learning, solving problems, and making decisions. AI is utilized to create systems and apps in higher education that can help increase the efficacy and efficiency of teaching and learning (Solís et al. 2023). The introduction of new technologies and teaching and learning methodologies, along with educational reforms, are the main drivers of these developments, according to EU policy (Bucea-Manea-Ţoniş et al. 2022). Though greatly impacted by other fields, artificial intelligence has its roots in computer science and engineering. A common definition and concept of artificial intelligence and intelligence in general are not widely agreed upon among scholars studying AI (Pedró 2022). Among the different forms of technology incorporated into education, artificial intelligence stands out as an important advancement (Slimi and Carballido 2023). AI application development is ingrained in its social framework. That implies that the way an AI application is created and functions is influenced by the norms, values, expertise, and attitudes of its developers (Sam and Olbrich 2023). Science fiction was once the domain of artificial intelligence, but today AI is a real, disruptive force in education. In an educational context, generative large language models (LLM)—like ChatGPT from OpenAI—perform and augment skills including data analysis, comprehension of intricate concepts, problem-solving, coding, and writing output production that typically need human intellect (Lacey and Smith 2023). Artificial intelligence and extended reality (XR; such as augmented reality and virtual reality) in education are essential steps toward constructivist and activity-based learning. Continue to be active and adapt to the needs of students and their educational paths, embedded with other resources like mobile devices for learning. Through the creative fusion of teaching and learning modalities, these tools generate innovative pedagogical approaches (Rangel-de Lázaro and Duart 2023).

The evolution of higher education has been profoundly influenced by the rapid integration of artificial intelligence and emerging technologies. This shift, driven by the need to align educational systems with market demands, has reshaped teaching, learning, and administrative processes. AI's impact spans various crucial areas within higher education, from improving student retention to equipping individuals with skills tailored for the dynamic job market of the fourth industrial revolution. Notably, AI's role extends beyond traditional instruction, encompassing Big Data management, personalized learning approaches, and the creation of more dynamic teaching methodologies. The COVID-19 pandemic catalyzed the adoption of technology in education, highlighting the collaborative potential between AI and extended reality (XR) tools in fostering student-centered learning environments, particularly in online education contexts. Education institutions are urged to cultivate new digital competencies, ranging from AI and machine learning (ML) to blockchain, VR/AR, and gamification, underscoring the necessity of adapting to these technological advancements. The European Union recognizes these technological innovations and educational reforms as pivotal drivers of change within the educational landscape.

While AI significantly shapes higher education, there's ongoing debate and exploration about its definition, societal implications, and the intertwining of AI development with social frameworks. The real-world impact of AI in education is highlighted by its ability to augment various cognitive skills and revolutionize learning approaches, aided by the integration of large language models like ChatGPT. These technologies, including XR, have been instrumental in fostering innovative pedagogical approaches and promoting constructivist, activity-based learning. In summary, the proliferation of AI in higher education represents a fundamental paradigm shift, emphasizing the imperative for institutions to embrace technological advancements, integrate them effectively into educational practices, and continually adapt to meet the evolving needs of students and society at large.

4.2 FOUNDATIONS OF AI INTEGRATION: UNDERSTANDING AI'S ROLE IN HIGHER EDUCATION

The technology known as artificial intelligence has come to revolutionize several industries, including education. Artificial intelligence has great promise for revolutionizing teaching and learning approaches, boosting student involvement, and enhancing overall academic results in higher education (Nagaraj et al. 2023). Artificial intelligence has become a commonplace notion and instrument in modern culture, playing a crucial role in daily activities. As a result, to develop successful global citizens, students' education should include a fundamental grasp and knowledge of AI (Southworth et al. 2023). In the 1950s, artificial intelligence emerged. A thinking machine is a type of computer that can autonomously or interactively carry out a variety of humanistic functions in a variety of situations. Turing first proposed the idea of a thinking machine in 1950 (Ge and Hu 2020). The use of AI in education, particularly in higher education, is becoming more popular and is being tested. AI is thought to advance education and bring in fresh approaches to teaching and learning (Slimi and Carballido 2023). Artificial intelligence technology advancements in recent years have fundamentally altered how people work, communicate, and learn. Higher education is one area where the potential advantages of AI could be very beneficial. Institutions and educators are researching the integration of AI applications to improve the educational process and student outcomes in response to the growing need for innovative teaching and learning approaches (Clita Pinto 2023). Application of Artificial intelligence in education, deep learning, and teaching assistants is all becoming more and more necessary as distance learning gains traction (Slimi and Carballido 2023). Machine learning, neural networks, AI-based data mining, and virtual reality are some of the AI technologies that are now in use. The future development direction of AI is expected to apply to the higher education sector, as evidenced by its inspiration from the human brain (Wang and Zhan 2021).

Neoliberal rationales, which further push for posthuman futures and technology to satisfy the efficiency and productivity goals of the neoliberal paradigm, are pervasive in the policy documents and portray AI as a solution to educational difficulties. Production of knowledge and business are closely related. Academics are both entrepreneurs and businesspeople at the same time and educational technology companies are fostered and grown by universities, while businesses also generate information (Gellai 2023). To predict future breakthroughs and the potential of new technologies, one must cultivate innovative and transversal abilities (Bucea-Manea-Ţoniş et al. 2022). High-quality instruction and learning outcomes, as well as the acquisition of relevant competencies, are the goals of HE. Appropriate techniques and approaches are employed in the assessment of educational quality (Bucea-Manea-Ţoniş et al. 2022). Higher education is one of the many industries that AI is changing. The pandemic has demonstrated the potential of AI to enhance teaching and learning in higher education (Slimi and Carballido 2023). Through the use of AI, decision-makers and administrators may carry out their duties more effectively and efficiently, which enhances the quality of education. This benefit extends beyond students and teachers (Salas-Pilco and Yang 2022).

AI creates pertinent theories and technologies while researching how to replicate and carry out some of the cognitive processes of the human brain in robots. In essence, the goal is to use computers to mimic human intelligence—that is, to replicate human thought, awareness, and behavior (Ge and Hu 2020). The old educational knowledge system will be subverted in the vast data environment by applying AI technology, resulting in an education that is continuously diversified, lacks boundaries between disciplines, and interacts with professional education (Ge and Hu 2020). AI is a new kind of research that combines Big Data, cloud computing, computer science, mobile internet, sensors, brain science, education neurology, bionics, social science, and other cutting-edge comprehensive fields. The field of artificial intelligence that heavily integrates and crosses over with educational technology, information technology, neural networks, educational science theory, etc. to support the intellectualization of management and education is known as educational AI (Ge and Hu 2020).

Because of their user-friendly chatbot interfaces, many generative artificial intelligence platforms are freely accessible. When utilized effectively, the outputs can significantly boost productivity since they are provided in a believable, human-like manner. The text generated by these models mimics human-like communication through the use of machine-learning algorithms. This feature has significant implications for educational procedures (Lacey and Smith 2023). AI integration in STEM higher education has already shown promise in improving research, teaching, and learning. Promising outcomes have been observed in the current uses of AI in data analysis, collaborative learning, intelligent tutoring, personalized learning, assessment, and feedback (Nagaraj et al. 2023; Rezaev and Tregubova 2023) sought to formulate a general problem, highlight theoretical and methodological foundations, and determine the logic of further research into the higher education system in the context of the widespread dissemination of AI technologies and the growing interdependence between humans and algorithms. The administration of education and instruction in colleges and universities has also seen new life and prospects with the development of AI. The individuation-based ecological system of educational AI is progressively taking shape. Through a variety of instructional activities, educators, students, and the learning environment are inextricably interwoven in the full dynamic system that is a college education (Ge and Hu 2020).

The widespread integration of artificial intelligence is reshaping higher education, promising transformative impacts on teaching, learning methodologies, and overall academic outcomes. AI's evolution, from a concept conceived in the 1950s to its current practical applications, signifies a significant shift in educational paradigms. The influence of AI in education extends beyond traditional instruction, encompassing various technologies such as machine learning, neural networks, and AI-based data mining. These innovations are heralded as pivotal tools to enhance educational processes, particularly in response to the demands of distance learning and the evolution of teaching methodologies. However, the integration of AI in education isn't without its complexities. Discussions revolve around the broader societal implications, including neoliberal influences that present AI as a solution to educational challenges. Nonetheless, AI's potential to enhance decision-making, administrative processes, and educational quality is evident, transcending benefits solely for students and teachers. AI's emergence is seen as a means to replicate human intelligence, altering educational systems within a vast data environment. This development heralds an education system characterized by continuous diversification, breaking boundaries between disciplines, and fostering interaction with professional education.

The field of educational AI, an interdisciplinary domain combining various cutting-edge fields, underscores the integration and crossover of technologies to support educational management and intellectualization. Notably, the accessibility and utility of AI platforms, including chatbots, hold significant promise for enhancing productivity and communication within educational procedures. Moreover, AI's integration into STEM education demonstrates promising outcomes in research, collaborative learning, personalized learning, and assessment techniques. In academia, research into the interdependence between humans and algorithms within the higher education system is gaining traction, highlighting the need to understand the implications and potentials of widespread AI integration. Ultimately, the developmental trajectory of educational AI signifies an individuation-based ecological system wherein educators, students, and the learning environment become dynamically interwoven, reshaping the dynamics of college education.

4.3 AI TOOLS AND IMPLEMENTATION: INNOVATIVE AI TOOLS FOR EDUCATION MANAGEMENT AND ITS APPLICATION

Our access to a wealth of information is made possible by the internet and Big Data technologies, which have significantly altered the social and economic landscape. As a result, there has been a lot of interest in their use in college and university teaching administration and education. With its more sophisticated, individualized, precise, and varied qualities, artificial intelligence offers a fresh perspective on how instructional management techniques could be innovated in colleges and universities (Ge and Hu 2020). With the use of various AI programs, academics and students may now easily produce essays, blogs, video transcripts, reflections, workflows, summarize peer-reviewed publications, and more. AI prompts that are well designed can educate users and provide them with pertinent, tailored insights on real-time data (Lacey and Smith 2023). AI can enhance assessment quality, offer useful learning and teaching strategies, and raise the caliber of education to better prepare students for the workforce. Slimi and Carballido (2023) highlight how AI has the potential to influence job prospects in the future and how higher education institutions must use AI to keep up with industry expectations. Applications of AI are highly beneficial for the growth of knowledge management, learning, teaching, and skill development (Slimi and Carballido 2023). The potential of artificial intelligence to expand and enhance educational opportunities has been hailed, making it a cybernetic (Gellai 2023).

Inclusionary approaches also benefit from AI, especially when considering students who are dyslexic or neurodivergent. Such students may find it difficult to comprehend lengthy, complicated written works; nevertheless, AI is capable of summarizing or rephrasing scientific literature in a way that is more helpful to the individual student (Lacey and Smith 2023). A dynamic system is also formed by the government, educational institutions, businesses, and centers for scientific research. The government supports colleges financially and via policies. Internet businesses progressively introduce AI technology into the educational space by depending on their benefits and features (Ge and Hu 2020). The application of artificial intelligence in education has a revolutionary impact on educational quality. Artificial intelligence in education has decreased the demand for academic staff and the maintenance of financially independent institutions while also helping learners communicate and connect with the outside world more effectively. It has been determined that incorporating cloud, IoT, and AI technologies into curricula for higher education can be beneficial for innovation and curriculum development (Slimi and Carballido 2023).

In comparison to other industries, the use of artificial intelligence in education and classroom management at colleges and universities is still in its infancy, despite some foundational successes (Ge and Hu 2020). It is how artificial intelligence can be used to improve teaching and learning in colleges and universities by actively creating the AI ecosystem of higher education and developing novel application modes from the perspectives of faculty, students, and teaching management. AI applications will raise the bar for management in the fields of education and instruction (Ge and Hu 2020).

Although integrating AI into education has the potential to lead to creative instructional approaches, this research may have been more thorough if they had assessed AI in a range of qualitative and quantitative scenarios (Slimi and Carballido 2023). Køhler Simonsen and Bidarra de Almeida (2020) examined how AI applications might support particular learning activities in higher education, analyzed and discussed the pedagogical applications of AI in higher education, and made the case that this particular decision support tool could assist educators in choosing the appropriate AI applications for the appropriate learning types and implementing the appropriate learning activities. Slimi and Carballido (2023) draw attention to an increasing amount of research that emphasizes the value of AI in higher education as well as the possible benefits of incorporating it into the curriculum. It is critical to recognize the biases and unfavorable effects that could result from integrating AI into the educational system.

The policy texts are infused with neoliberal rationales that further promote posthuman futures and technology to satisfy the efficiency and productivity requirements of the neoliberal paradigm. AI is positioned as a solution to educational difficulties on these grounds. The production of information is intricately linked to business (Gellai 2023). Incubators and accelerators of educational technology enterprises, universities themselves are knowledge creators, and academics are both entrepreneurs and businesspeople (Gellai 2023). The goal of the study is to illustrate the value of artificial intelligence and provide instances of AI-based solutions for improving (creating, assessing, and overseeing) customer engagement (CE) on social media in the higher education sector. One of the most important non-financial metrics for business performance in digital marketing strategy nowadays is CE. The paper acknowledges the important role AI plays in boosting CE in social media and proposes a decision support system (DSS) based on social media engagement management with the use of AI-based tools in a case study of the higher education business. It was made quite clear that colleges may operate more efficiently and improve their non-financial performance by using AI in marketing (Golab-Andrzejak 2022). Artificial intelligence-based technologies surely serve universities in developing, assessing, and monitoring student involvement. This aids in decision-making and makes AI-based tools a permanent part of the decision support system (Golab-Andrzejak 2022).

Robotics and artificial intelligence could have a significant long-term impact on colleges. The fragmented character of the pertinent literature and the emphasis on the social, ethical, pedagogical, and administrative issues surrounding automation through AI and robots in higher education make it difficult to fully comprehend the extent of this impact (Naqvi et al. 2023).

Artificial intelligence has garnered significant scientific attention in the past ten years across a wide range of industries, including finance, law, and medicine. AI's promise in education has led to a recent focus on its application in this field of study (Salas-Pilco and Yang 2022). Predictive modeling, intelligent analytics, assistive technology, autonomous content analysis, and image analytics are the primary uses of AI in education. AI applications support the provision of high-quality education by addressing significant difficulties in the field of education (Salas-Pilco and Yang 2022).

The University of Florida (UF) is integrating artificial intelligence into all parts of the curriculum and creating chances for student participation in specific AI literacy domains, irrespective of the academic program. All students will have access to AI education as a fundamental opportunity according to the UF program, AI Across the Curriculum. The ultimate objective of AI Across the Curriculum is to develop a workforce that is prepared for the twenty-first century and possesses the fundamental skills that are required by governments and the workforce around the globe (Southworth et al. 2023). To address the challenges of the twenty-first century, qualified human capital is crucial, and UF is putting itself in a position to lead in supplying this demand on society at large. A range of AI possibilities are offered to all students as part of the AI Across the Curriculum paradigm, and their participation is highly encouraged (Southworth et al. 2023).

The integration of artificial intelligence into higher education is poised to revolutionize teaching methodologies, administrative processes, and student learning experiences. AI, with its sophisticated and precise capabilities, presents a fresh perspective on instructional management techniques in colleges and universities. Its utilization spans various applications, enabling easy creation of essays, blogs, video transcripts, and personalized insights through AI prompts. Additionally, AI enhances assessment quality, teaching strategies, and student preparation for the workforce, influencing job prospects in the future. AI's inclusive approaches benefit students with learning differences by summarizing complex material in accessible ways. The collaborative dynamics between the government, educational institutions, businesses, and research centers foster AI's integration into the educational space. Despite being in its early stages in comparison to other industries, AI in education is rapidly evolving, promising improved management and innovative teaching and learning experiences. However, there are concerns about potential biases and adverse effects resulting from AI integration. Neoliberal influences often position AI as a solution to educational challenges, intertwining information production with business motives. Research underscores AI's value in education while cautioning against biases and unfavorable implications. Studies explore AI's role in decision support systems, customer engagement enhancement in social media, and its impact on student involvement and non-financial performance in higher education. Initiatives like the University of Florida's AI Across the Curriculum program aim to democratize AI education for all students, preparing them for the demands of the twenty-first-century workforce. The multifaceted impact of AI on education encompasses predictive modeling, analytics, assistive technology, and content analysis. However, the extensive implications of robotics and AI in higher education, encompassing social, ethical, pedagogical, and administrative aspects, remain complex and fragmented in current literature. Overall, AI's promise in education is evident across various applications and initiatives, signaling a transformative shift in preparing students for future challenges and opportunities in a rapidly evolving global landscape.

4.4 CHALLENGES AND ETHICAL CONSIDERATIONS OF USING AI IN HE

Though AI-powered technologies are becoming more and more common in many fields, including education, there is little research on AI in higher education governance. However, there are many theories and reasons supporting or opposing the adoption of these technologies because they are still in their infancy and have not yet realized their full potential (Gellai 2023). AI is giving rise to a lot of ethical problems. In the future AI society, professionals who study and create AI technology will require ethical education. This is mostly because of their ethical skills, which produce ethical AI products and services (Lim, Seo, and Kwon 2022). The difficulties associated with AI adoption at HEIs have an impact on AI tool actions. Digital illiteracy and the integration of cognitive efforts with system and privacy issues are factors that have higher loading factors and importance when it comes to AI implementation issues. These issues are essential barriers to AI. Furthermore, a major obstacle to AI is the lack of knowledge of AI technology. These components seem to motivate academics to learn new cross-disciplinary AI skills and knowledge, which has a positive impact on the HEI system (Bucea-Manea-Ţoniş et al. 2022).

One could argue that artificial intelligence in education (AIED) is the next major force to upend higher learning. There is very little real-world experience with AI in higher education, and little thought has been devoted to the technology's possible pedagogical uses (Køhler Simonsen and Bidarra de Almeida 2020). Artificial intelligence is finding ever-greater uses in education, but widespread use of AI still seems a long way off. Though artificial intelligence presents vast potential to enhance instruction and learning, creating applications for higher education entails several ethical and practical concerns (Pedró 2022). Administrators would be inclined to substitute affordable automated AI solutions for instruction during budget cuts during a catastrophe. Intelligent tutors, expert systems, and chatbots may replace faculty, teaching assistants, educational advisers, and administrative staff—and maybe for good cause. Serious privacy and data protection concerns are raised by the huge volumes of data needed for the use of AI, particularly in learning analytics. This data includes private student and teacher information (Pedró 2022). The use of AI in higher education is not without its difficulties and worries. There are worries that excessive reliance on technology and automation may lower human contact and educational standards (Solís et al. 2023).

The development of artificial intelligence has the potential to widen the already-existing gaps in access to and involvement in higher education, dividing the population between those with and without access to these technologies (Solís et al. 2023). Benefits of AI include personalized recommendations, customized feedback, and learning that is more individualized. These findings have been shown to raise the caliber of higher education. The ability to customize lesson plans and match the unique needs of each student is made possible by technological advancements, which enhance the quality of education kids receive overall (Solís et al. 2023). Like with any significant invention in life, it's critical to comprehend the moral conundrums raised by this. Many believe that humans need interpersonal relationships to support and guide their learning process, and many fear that AI task automation will remove or replace occupations. AI cannot provide the social and emotional support that people require in a learning setting (Crompton and Song 2021). The increasing prevalence of academic dishonesty (AD) in which students use artificial intelligence or commercial essay writing services (essay mills) jeopardizes the legitimacy of evaluation methods used in higher education (HE) across the globe (Sweeney 2023). To better understand why students fall victim to essay mills and other similar services, universities should involve their students in the process. Improved learner education is essential to helping students comprehend academic misconduct and the dangers it brings to their future academic and professional opportunities (Sweeney 2023). Universities must give their faculty members excellent in-service training. This has to involve education on spotting and handling academic dishonesty as well as the introduction of AI tools like ChatGPT and Bard. While more conventional contract cheating may be disrupted by AI, innovative and imaginative forms of evaluation ought to be a part of the solution (Sweeney 2023).

Higher education is increasingly utilizing artificial intelligence or AI. This increase has raised concerns about the morality of AI systems. As a result, it is important to comprehend how the concepts of FATE—Fairness, Accountability, Transparency, and Ethics—are defined in the AI and higher education research that has been done thus far (Memarian and Doleck 2023). It is discovered that the term "fairness" is the most frequently used and researched in the reviewed studies and that it has a more technical meaning. The dearth of research on transparency and accountability, the all-encompassing nature of ethics, and the hazy definitions of bias and fairness call for further discussion and clarification (Memarian and Doleck 2023). AI-driven surveillance technologies have made a significant impact on the university student experience in North America, transforming everything from learning materials to assignments and exams. "Technosaviorism" is the term used to describe the adoption of these tools, which save time and help catch plagiarism more quickly while also better-serving students (Swartz and McElroy 2023). Students are left with no choice but to accept the risks when these surveillance technologies are integrated into the prerequisites for enrollment in degree-granting courses, even though they can take certain precautions to protect their privacy (Swartz and McElroy 2023).

The emergence of Artificial Intelligence in higher education is raising profound questions and considerations regarding its ethical, pedagogical, and practical implications. While its potential for enhancing instruction and learning is vast, the integration of AI presents complex challenges. Concerns surrounding AI in higher education span various dimensions. Ethical issues, including data privacy and the potential replacement of human educators with automated systems, provoke discussions about the moral conundrums AI brings. Additionally, academic dishonesty through AI or essay mills threatens evaluation methods, demanding universities to engage students in understanding and combating academic misconduct. AI's capacity for personalized learning experiences, such as tailored feedback and individualized lesson plans, highlights its potential to enhance education quality. However, doubts persist regarding AI's ability to provide the essential social and emotional support crucial for the learning process. Moreover, AI's implementation in higher education governance remains in its infancy, prompting discussions about digital illiteracy, knowledge gaps, and the need for interdisciplinary AI skills among educators. The concepts of fairness, accountability, transparency, and ethics in AI systems within higher education are under scrutiny, with a specific emphasis on the technicalities of "fairness" and the need for further research on transparency and accountability. Surveillance technologies powered by AI are transforming the student experience but also raise concerns about privacy and acceptance of risks. These technologies, while promising time-saving features and efficient plagiarism detection, pose dilemmas for students concerning privacy infringement. In summary, the growing presence of AI in higher education presents a dichotomy: while offering promising avenues for personalized learning and efficiency improvements, it also sparks multifaceted debates regarding ethics, privacy, and the role of human educators. Further exploration and discussions are imperative to navigate these complex implications and ensure the ethical and effective integration of AI in higher education.

4.5 MITIGATING CHALLENGES: STRATEGIES FOR SUCCESSFUL AI INTEGRATION IN HIGHER EDUCATION

There are advantages and disadvantages to integrating artificial intelligence into higher education. The quality and accessibility of education can be enhanced by AI, but there are social and ethical ramifications that must be considered as well. It is important to put into practice policies and tactics that, in the context of AI in higher education, support equity and interpersonal relationships. To optimize AI's positive effects and minimize any potential drawbacks, it is also necessary to continuously consider the technology's position in education (Solís et al. 2023). Comprehending the sociological aspects of the difficulties is crucial to comprehend the pedagogical and sociocultural consequences of artificial intelligence in higher education. This entails investigating how the use of AI impacts social dynamics in the classroom, power dynamics in educational institutions, and student-teacher interactions (Solís et al. 2023). Ethical domain specialists who can assess societal shifts, identify moral dilemmas, and carry out morally sound solutions to address these problems are essential for the ethical AI society. It is very difficult for the general public to learn about and regulate the immoral behavior of highly specialized professionals, like AI developers, because they have a practical monopoly on information and technical innovation in their field. Thus, the AI developer needs a high level of morals and ethics commensurate with their expertise and independence in their field (Lim, Seo, and Kwon 2022).

In addition to teaching competent developers how to write code, universities ought also to empower their students to assume accountability for whatever problems their work may cause in the AI community. We must create practical ethics education if we are to succeed in this endeavor (Lim, Seo, and Kwon 2022). Teaching quality, student management, faculty construction, and teaching management at an intelligent level will all continue to improve with the ongoing progress of AI technology and its extensive application in university administration (Ge and Hu 2020). Enhancing non-experts' AI skills is crucial and will only grow more so in the future, as artificial intelligence permeates more and more aspects of daily life. Although young children must be exposed to the potential benefits of artificial intelligence, adults pursuing higher education and beyond should also possess a fundamental grasp of AI, or AI literacy, to properly engage with the technology (Laupichler et al. 2022).

Artificial intelligence has a significant impact on both our personal and professional lives. University students must gain a fundamental understanding of artificial intelligence to navigate the difficulties and recognize the potential presented by this technology (sometimes referred to as "AI literacy"). The degree of students' AI literacy must be determined to create study plans that effectively promote AI competencies. Effective AI courses are required for a wider range of students who use AI in their daily lives and will utilize AI tools in their future employment. When creating AI courses, educators should take into account the variation in students' prior knowledge (Hornberger, Bewersdorff, and Nerdel 2023). The integration of artificial intelligence in higher education presents a dual landscape of opportunities and challenges. While AI holds the potential to enhance the quality and accessibility of education, it also introduces social, ethical, and pedagogical considerations that demand scrutiny.

Acknowledging the multifaceted impact of AI in education, it becomes imperative to implement policies and strategies fostering equity and interpersonal relationships within the AI framework. This involves a continuous assessment of AI's role and position in education to optimize its benefits while mitigating potential drawbacks. Understanding the sociological aspects of AI's integration is crucial, encompassing examinations of social dynamics in classrooms, power structures within educational institutions, and the dynamics of student-teacher interactions. Ethical considerations emerge prominently, emphasizing the need for specialists capable of evaluating societal shifts, identifying moral dilemmas, and devising ethical solutions. Ethics education becomes essential not just for technical AI developers but also for non-experts who interact with AI in various aspects of their lives.

Furthermore, the progression of AI technology in university administration signifies ongoing improvements in teaching quality, student management, faculty construction, and educational management at an intelligent level. Recognizing the growing significance of AI literacy, both among young children and adults pursuing higher education, underscores the importance of cultivating a fundamental understanding of AI. Curricular planning needs to account for students' varying levels of AI knowledge, offering effective AI courses that cater to a broad spectrum of learners with diverse prior knowledge. In essence, the integration of AI in higher education necessitates a holistic approach that not only harnesses its potential for educational enhancement but also navigates the ethical and societal implications, emphasizing the need for education and ethical frameworks to align with this transformative technology.

4.6 FUTURE DIRECTIONS AND OPPORTUNITIES: THE ROLE OF AI IN SHAPING FUTURE EDUCATIONAL LANDSCAPES

Anticipating future technology potential and breakthroughs requires the development of creative and transversal talents. The emphasis of the new educational approaches is on morals, values, problem-solving, and day-to-day activities. Future educational reforms must take into account the content that students are studying, their potential to acquire crucial skills, and their learning style (Bucea-Manea-Ţoniş et al. 2022). It is yet difficult to predict the full effects of AI application development in higher education, but it is quite likely that these issues will persist for the next 20 years. Thus far, apps have been developed that support student learning in personalized and adaptable learning environments and present a wealth of pedagogical potential for intelligent student support systems (Pedró 2022). Although it is yet difficult to predict with certainty, the overall effects of AI application development in higher education are expected to be a major concern over the next 20 years. Personalized and adaptive learning environments are supported by the apps that have been built thus far, providing vast pedagogical prospects for intelligent student support systems (Pedró 2022). AI has the potential to alleviate the challenge of accommodating a growing student body's demand for higher education, particularly at open and remote learning colleges (Pedró 2022). The application of AI in higher education has created new opportunities for learning personalization. Learning materials and activities can be customized to each student's needs thanks to AI systems' ability to gather and analyze vast volumes of student data. This can boost teaching procedures' effectiveness and greatly enhance the educational experience (Solís et al. 2023). Thorough attention must be paid to issues related to algorithmic accountability, automated decision-making, and student data privacy. Furthermore, it's critical to think about how less human connection affects learning and how it can impact students' development of social skills (Solís et al. 2023).

In the future, evaluating student prompt engineering rather than the AI-written product could be one way to help them delve deeper into a subject and produce more meaningful results. Then, with the aid of AI techniques, students can comprehend difficult subjects by receiving enhanced notes on the course contents. Through advancements in spelling, grammar, and structure, the efficient application of AI can boost student productivity and produce work of a high caliber (Lacey and Smith 2023). To actively create the AI ecosystem of higher education and innovate the application mode from the perspectives of student management, faculty construction, and teaching management, this is the direction in which AI may be applied to the management of education and teaching in colleges and universities. The management of education and teaching will go to a new level with the use of AI (Ge and Hu 2020). Educators could utilize the picker wheel approach to AIED decision assistance tool in practice to help with pedagogically sound decisions while developing curricula in higher education. Indeed, learning activities and artificial intelligence are a marriage made in heaven, but educators should guide AI, not follow it (Køhler Simonsen and Bidarra de Almeida 2020).

Implementing sustainable, ethical governance is imperative as the need for artificial intelligence grows worldwide. Both developed and developing countries now understand how important it is to put ethical AI usage rules and guidelines into place (Slimi and Carballido 2023). Great benefits will be preserved for such institutions by developing possible future steps, getting instructions and guidance for applying new technologies in HEI, and resolving any potential challenges in the adoption and implementation of these technologies (Bucea-Manea-Ţoniş et al. 2022). Digital technologies provide a plethora of opportunities and ways to facilitate the transformation of education; therefore, it is imperative to adopt a comprehensive perspective on technology-enhanced learning, recognize the necessary adjustments, and comprehend its impacts (Bucea-Manea-Ţoniş et al. 2022).

While artificial intelligence is a major engine behind VR and AR, AI technology limits the variety of interactions. We are just beginning to see how AI and VR can be combined to advance the field of higher education. AI's development in higher education is probably going to be constrained by its flaws. Thus, from the standpoint of advancing technology, it is imperative to further integrate AI with higher education (Wang and Zhan 2021). Virtual reality and artificial intelligence are two areas that are showing signs of convergence and intersection, which significantly increases the level of involvement in education (Wang and Zhan 2021). In the areas of IA, machine learning, IoT, 5G, cloud computing, Big Data, blockchain, and data analysis, colleges need to develop new digital skills. These include using MS Office and other apps, MOOCs, simulation software, VR/AR, and gamification. They also need to cultivate a long-term perspective and cross-disciplinary abilities (Bucea-Manea-Țoniş et al. 2022).

Future developments in AI will concentrate on developing platforms for adaptive learning, promoting online and distant learning, expanding scientific research, and tackling ethical issues. Higher education in STEM fields can better educate students about the problems of the future and encourage innovation in scientific research and technological developments by utilizing AI (Nagaraj et al. 2023). AI in STEM higher education has enormous potential to have a revolutionary impact in the future. As AI technologies improve, more research and development should be focused on a few important areas (Nagaraj et al. 2023). First off, the demand for adaptive learning platforms driven by AI that can smoothly interface with current learning management systems (LMS) and educational platforms is rising. These platforms ought to be able to monitor student progress, deliver individualized learning experiences, and make astute suggestions for additional readings or other resources (Nagaraj et al. 2023; Drach et al. 2023) systematically organizes the ethical standards that should be followed when utilizing AI, including human-centered values, governance, accountability, transparency, and sustainability. The SWOT analysis assisted in identifying the benefits, drawbacks, possibilities, and dangers associated with implementing AI in higher education. People with physical disabilities may benefit from artificial intelligence, which has applications in a variety of fields, including the automobile sector. A higher standard of living may be used in government operations to reduce workloads and streamline processes. While it is possible to be outwitted, this risk can be reduced with sufficient supervision. It increases unemployment even if it might increase workplace productivity (Clita Pinto 2023).

AI technologies have the potential to enhance the inclusion, accessibility, and effectiveness of present teaching strategies. However, addressing and resolving AI-related issues in higher education is crucial. The main obstacles that must be properly addressed are those related to privacy and data security, ethical concerns, and the influence on the interactions between educators and students. It's also critical to pay attention to the expanding problem of knowledge transfer that AI applications are affecting (Clita Pinto 2023). The integration of artificial intelligence and extended reality (XR), such as augmented reality (AR) and virtual reality (VR), into education is a crucial step toward constructivist and activity-based learning (Rangel-de Lázaro and Duart 2023). Integrated learning tools, like mobile devices, are always evolving to meet the needs of students and fit their unique educational paths. With the use of these technologies, innovative pedagogical approaches are created, combining creative teaching and learning modalities (Rangel-de Lázaro and Duart 2023). The integration of artificial intelligence in higher education holds substantial promise for transforming learning methodologies and educational paradigms. However, it also presents various challenges that require strategic consideration and ethical frameworks. AI applications offer extensive opportunities for personalized and adaptive learning environments. They enable customized learning materials and activities, catered to individual student needs by leveraging vast data analysis capabilities. These advancements enhance teaching effectiveness and significantly elevate the overall educational experience. While the potential for AI-driven learning is vast, critical attention is demanded in several domains. Algorithmic accountability, automated decision-making, student data privacy, and the impact of reduced human connection on learning and social skills development are crucial areas necessitating scrutiny.

The future direction of AI in education spans multiple fronts. Educators need to guide AI integration, ensuring pedagogically sound decisions while leveraging AI's capabilities. Efforts must be concentrated on developing platforms for adaptive learning, promoting online and distance education, and addressing ethical concerns within AI's purview. The convergence of AI with other technologies like virtual reality (VR) and augmented reality (AR) offers further possibilities for constructivist and activity-based learning. However, the integration of AI with higher education demands comprehensive approaches, encompassing technological and ethical considerations, emphasizing the need for balanced progress and ethical governance. As AI continues to evolve and intertwine with education, ongoing initiatives in education must acknowledge the ethical, privacy, and social interaction aspects of AI adoption. This involves cultivating AI literacy among students, educators, and developers and embracing a proactive stance toward technology-enhanced learning while balancing it with human-centric pedagogies and ethical frameworks.

4.7 CONCLUSION

The integration of artificial intelligence into higher education represents a pivotal juncture that promises transformative advancements while concurrently raising complex considerations and challenges. Across diverse discussions, it's evident that AI's infusion has redefined educational paradigms, reshaping teaching methodologies, learning approaches, and administrative processes. The multifaceted impact spans from personalized learning and improved student outcomes to the cultivation of digital competencies essential for the evolving job market. However, this evolution is not devoid of complexities. Debates revolve around ethical, societal, and pedagogical implications, as well as concerns about potential biases, data privacy, and the balance between technological advancements and human-centered education. The convergence of AI with emerging technologies like extended reality (XR) offers vast potential for constructivist learning but necessitates a holistic approach encompassing technological sophistication and ethical governance. The ongoing exploration of AI's role in education emphasizes the necessity for continuous assessment, sociological understanding, and ethical frameworks to optimize its benefits and mitigate potential drawbacks. AI literacy emerges as a crucial facet, demanding educational systems to accommodate varying levels of AI knowledge and cultivate comprehensive AI education across diverse learner backgrounds. In conclusion, the integration of AI in higher education heralds a fundamental shift that necessitates a harmonious blend of technological innovation and ethical stewardship. Navigating the complexities and harnessing the full potential of AI in education demands a proactive, adaptive approach that aligns educational practices with the evolving needs of students and society, ensuring that the transformative power of AI is wielded responsibly and inclusively.

REFERENCES

Bucea-Manea-Ţoniş, Rocsana, Valentin Kuleto, Simona Corina Dobre Gudei, Costin Lianu, Cosmin Lianu, Milena P. Ilić, and Dan Păun. 2022. "Artificial Intelligence Potential in Higher Education Institutions Enhanced Learning Environment in Romania and Serbia." *Sustainability (Switzerland)* 14 (10). https://doi.org/10.3390/su14105842

Clita Pinto, Tressy 2023. "Artificial Intelligence (AI) Applications in Higher Education: Addressing the Elephant in the Room: Case Study on Ai." *EPRA International Journal of Environmental Economics, Commerce and Educational Management*, no. September: 16–20. https://doi.org/10.36713/epra14282

Crompton, Helen, and Donggil Song. 2021. "The Potential of Artificial Intelligence in Higher Education." *Revista Virtual Universidad Católica Del Norte* 6 (62): 1–4. https://doi.org/10.35575/rvucn.n62a1

Drach, Iryna, Olha Petroye, Oleksandra Borodiyenko, Iryna Reheilo, Oleksandr Bazeliuk, Nataliia Bazeliuk, and Olena Slobodianiuk. 2023. "The Use of Artificial Intelligence in Higher Education." *International Scientific Journal of Universities and Leadership* 15. https://doi.org/10.31874/2520-6702-2023-15-66-82

Ge, Zhenxing, and Ying Hu. 2020. "Innovative Application of Artificial Intelligence (AI) in the Management of Higher Education and Teaching." *Journal of Physics: Conference Series* 1533 (3): 1–6. https://doi.org/10.1088/1742-6596/1533/3/032089

Gellai, Dániel Béla. 2023. "Enterprising Academics: Heterarchical Policy Networks for Artificial Intelligence in British Higher Education." *ECNU Review of Education* 6 (4): 568–596. https://doi.org/10.1177/20965311221143798

Golab-Andrzejak, Edyta. 2022. "Enhancing Customer Engagement in Social Media with AI - a Higher Education Case Study." *Procedia Computer Science* 207: 3022–3031. https://doi.org/10.1016/j.procs.2022.09.361

Hornberger, Marie, Arne Bewersdorff, and Claudia Nerdel. 2023. "What Do University Students Know about Artificial Intelligence? Development and Validation of an AI Literacy Test." *Computers and Education: Artificial Intelligence* 5 (September): 1–12. https://doi.org/10.1016/j.caeai.2023.100165

Køhler Simonsen, H., and J. M. E Bidarra de Almeida. 2020. "Enhancing the Human Experience of Learning with Technology: New Challenges for Artificial Intelligence And Learning Activities: A Match Made In Heaven?" In *European Distance and E-Learning Network (EDEN) Proceedings 2020 Research Workshop | Lisbon*, 2707–2819. https://doi.org/10.38069/edenconf-2020-rw0022

Lacey, Melissa M., and David P. Smith. 2023. "Teaching and Assessment of the Future Today: Higher Education and AI." *Microbiology Australia* 44 (3): 124–126. https://doi.org/10.1071/MA23036

Laupichler, Matthias Carl, Alexandra Aster, Jana Schirch, and Tobias Raupach. 2022. "Artificial Intelligence Literacy in Higher and Adult Education: A Scoping Literature Review." *Computers and Education: Artificial Intelligence* 3 (May): 100101. https://doi.org/10.1016/j.caeai.2022.100101

Lim, Ji Hun, Jeong Eun Seo, and Hun Yeong Kwon. 2022. "The Role of Higher Education for the Ethical AI Society." In *Proceedings of the International Florida Artificial Intelligence Research Society Conference, FLAIRS*. Vol. 35. https://doi.org/10.32473/flairs.v35i.130609

Memarian, Bahar, and Tenzin Doleck. 2023. "Fairness, Accountability, Transparency, and Ethics (FATE) in Artificial Intelligence (AI) and Higher Education: A Systematic Review." *Computers and Education: Artificial Intelligence*. https://doi.org/10.1016/j.caeai.2023.100152

Nagaraj, B. K., A. Kalaivani, R. Suraj Begum, S. Akila, H. K. Sachdev, and N. Senthil Kumar. 2023. "The Emerging Role of Artificial Intelligence in STEM Higher Education: A Critical Review." *International Research Journal of Multidisciplinary Technovation* 5 (5): 1–19. https://doi.org/10.54392/irjmt2351

Naqvi, Syed Gulfraz, Faisal Iqbal, Javaria Yousaf, and Rimsha Tariq. 2023. "The Impact of Artificial Intelligence (AI) and Robotics on Higher Education." *Journal of Management Practices, Humanities and Social Sciences* 7 (3). https://doi.org/10.33152/jmphss-7.3.2

Pedró, Francesc. 2022. "Applications of Artificial Intelligence to Higher Education: Possibilities, Evidence, and Challenges." *IUL Research* 1 (1): 61–76. https://doi.org/10.57568/iulres.v1i1.43

Rangel-de Lázaro, Gizéh, and Josep M. Duart. 2023. "You Can Handle, You Can Teach It: Systematic Review on the Use of Extended Reality and Artificial Intelligence Technologies for Online Higher Education." *Sustainability* 15 (4): 1–23. https://doi.org/10.3390/su15043507

Rezaev, Andrey V., and Natalia D. Tregubova. 2023. "ChatGPT and AI in the Universities: An Introduction to the Near Future." *Vysshee Obrazovanie v Rossii* 32 (6): 19–37. https://doi.org/10.31992/0869-3617-2023-32-6-19-37

Salas-Pilco, Sdenka Zobeida, and Yuqin Yang. 2022. "Artificial Intelligence Applications in Latin American Higher Education: A Systematic Review." *International Journal of Educational Technology in Higher Education* 19 (21): 1–20. https://doi.org/10.1186/s41239-022-00326-w

Sam, Abraham Kuuku, and Philipp Olbrich. 2023. *The Need for AI Ethics in Higher Education*. SpringerBriefs in Ethics. https://doi.org/10.1007/978-3-031-23035-6_1

Slimi, Zouhaier, and Beatriz Villarejo Carballido. 2023. "Systematic Review: AI's Impact on Higher Education - Learning, Teaching, and Career Opportunities." *TEM Journal*. UIKTEN - Association for Information Communication Technology Education and Science. https://doi.org/10.18421/TEM123-44

Solís, MCS Wilton Melvin Villamar, Cecilia Alejandra García Ríos, Carlos Eduardo Cevallos Hermida, Jean Luis Arana Alencastre, and José-Horacio Tovalin-Ahumada. 2023. "The Impact of Artificial Intelligence on Higher Education: A Sociological Perspective." *Journal of Namibian Studies: History Politics Culture* 33. https://doi.org/10.59670/jns.v33i.969

Southworth, Jane, Kati Migliaccio, Joe Glover, Ja'Net N. Glover, David Reed, Christopher McCarty, Joel Brendemuhl, and Aaron Thomas. 2023. "Developing a Model for AI Across the Curriculum: Transforming the Higher Education Landscape via Innovation in AI Literacy." *Computers and Education: Artificial Intelligence* 4 (January): 1–10. https://doi.org/10.1016/j.caeai.2023.100127

Swartz, Mark, and Kelly McElroy. 2023. "The 'Academicon': AI and Surveillance in Higher Education." *Surveillance and Society* 21 (3). https://doi.org/10.24908/ss.v21i3.16105

Sweeney, Simon. 2023. "Who Wrote This? Essay Mills and Assessment – Considerations Regarding Contract Cheating and AI in Higher Education." *International Journal of Management Education* 21 (2). https://doi.org/10.1016/j.ijme.2023.100818

Wang, Ji, and Qinglong Zhan. 2021. "Visualization Analysis of Artificial Intelligence Technology in Higher Education Based on SSCI and SCI Journals from 2009 to 2019." *International Journal of Emerging Technologies in Learning* 16 (8): 20–33. https://doi.org/10.3991/ijet.v16i08.18447

5 The Future of Higher Education

Using AI in Universities to Improve Learning Outcomes and Operational Efficiency

Tarun Kumar Vashishth, Vikas Sharma, Kewal Krishan Sharma, and Bhupendra Kumar
IIMT University, Meerut, India

5.1 INTRODUCTION

The landscape of higher education is enduring a profound conversion with combination of AI, heralding a future where educational institutions leverage advanced technologies to enhance learning outcomes and operational efficiency. This chapter investigates into multifaceted impression of AI within university settings, exploring how these technologies can revolutionize both academic and administrative realms. AI presents a unique opportunity for universities to tailor learning experiences through personalized approaches, fostering increased student assignation and, consequently, enhanced learning outcomes. Concurrently, the adoption of AI-driven analytics offers invaluable insights into student performance, enabling timely interventions and customized support mechanisms. Beyond academics, the automation of routine administrative tasks through AI streamlines operational processes, allowing universities to allocate resources more efficiently. As universities embrace this transformative journey, they also grapple with ethical considerations, data privacy concerns, and the evolving role of educators. This overview sets stage of an all-inclusive examination of the opportunities and challenges associated with integrating AI in academics, offering insights into future landscape of academia (see Figure 5.1).

5.1.1 Background

The background of this chapter is rooted in the dynamic evolution of academics and increasing mixing of artificial intelligence with academic and administrative spheres. Traditional higher education models are facing unprecedented challenges and opportunities in the digital age, prompting universities to explore innovative solutions. The emergence of AI technologies provides a promising avenue to address these challenges and revolutionize the educational landscape. The backdrop is characterized

DOI: 10.1201/9781032644509-5

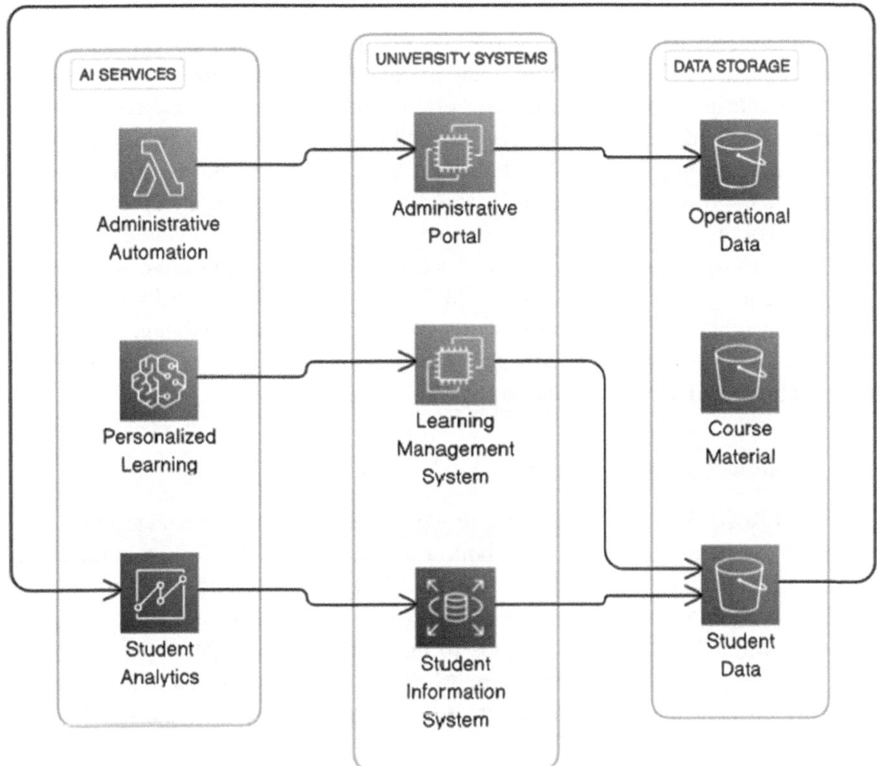

FIGURE 5.1 Higher Education AI Integration.

Source: Author created.

by a growing recognition that AI provides a pivotal role in reshaping learning experiences, adapting to diverse student needs, and optimizing administrative processes. Universities are motivated to harness the power of AI to expand learning outcomes, enhance student engagement, and streamline operational efficiency. The background also considers the broader context of technological advancements, data analytics, and automation, which are reshaping industries worldwide. As educational institutions navigate this transformative era, understanding the background of AI integration in higher education becomes crucial for envisioning upcoming trajectory of academic institutions and their pivotal role in shaping the next generation of learners.

5.1.2 Objectives of the Chapter

Primary aims of this chapter are to comprehensively explore and analyze the impact of artificial intelligence (AI) on the landscape of higher education. Firstly, the chapter aims to delve into transformative power of AI in improving learning outcomes within universities. By assessing the capabilities of AI in personalizing learning experiences and enhancing student engagement, the objective is to provide insights into how AI can positively influence the educational journey of students. Secondly, the chapter seeks to investigate the operational efficiency gains that AI can bring to higher education institutions. This includes a detailed examination of how AI-driven analytics and intelligent systems can optimize administrative processes, automate routine tasks, and facilitate resource allocation more effectively. Additionally, the chapter aims to shed light on the challenges and considerations associated with the integration of AI in higher education, including ethical concerns, data privacy issues, and the evolving role of educators. Furthermore, the objectives include identifying opportunities for future of AI in higher education, exploring innovative applications, and anticipating developments in the field. Ultimately, the chapter seeks to contribute valuable insights that inform stakeholders about the potentials, challenges, and future directions of leveraging AI to enhance both the educational and operational dimensions of higher education institutions.

5.1.3 Scope and Significance

This study's scope encompasses a thorough examination of the integration of AI in higher education, encompassing both academic and administrative dimensions. Study will delve into the transformative impact of AI on learning outcomes, exploring its potential to personalize educational experiences and enhance student engagement. Additionally, the scope extends to the assessment of operational efficiency improvements through the adoption of AI-driven analytics and intelligent systems, focusing on their ability to streamline administrative tasks and optimize resource allocation.

Significance of this study lies in its contribution to the broader understanding of how AI technologies can shape the future of higher education. By addressing specific challenges and opportunities within the academic and administrative realms, the study aims to provide actionable insights for educational institutions, policymakers, and educators. Understanding the ethical considerations, data privacy concerns, and the evolving role of educators in the context of AI integration is crucial for fostering responsible and effective implementation. Ultimately, the study's findings aim to guide stakeholders in harnessing the benefits of AI to create more adaptive, personalized, and efficient higher education ecosystems, thereby shaping the trajectory of academic institutions in the digital age.

5.2 LITERATURE REVIEW

In Hannan, Erin, and Shuguang Liu (2023) discuss how incorporation of AI technology in higher studies has become a key factor in maintaining competitiveness for universities around the world. The authors highlight profits and challenges of incorporating AI in higher education and provide recommendations for universities to effectively utilize this technology. In George, Babu, and Ontario Wooden (2023) delved into power impact of AI on the future of higher studies and how institutions can effectively manage this transformation. The authors emphasized the need for collaboration and innovation in order to adapt to the changing landscape of education. In Kuleto et al. (2021) highlight the potential of AI and ML in higher education, such as personalized learning, data analysis, and automation of administrative tasks. They also discuss the challenges that need to be addressed, including ethical concerns, data privacy, and the need for proper training and resources for faculty and staff. In Pedró, Francesc (2020) discusses the strength of AI in transforming the traditional methods of teaching and learning in academics of higher studies. With the increasing use of technology, AI has the ability to personalize learning experiences for students, provide real-time feedback, and improve overall efficiency in education. The chapter also highlights the evidence of successful implementation of AI in some universities and educational institutions. From chatbots to virtual tutors, AI has been used to enhance student engagement, improve retention rates, and provide personalized support to students. In Satya Vir Singh and Kamal Kant Hiran (2022) explore the potential effects of AI on the traditional methods of teaching and learning in higher education. In Muhie, Amedie, and Woldie (2020) explore mixing of AI in teaching and learning in higher education and its strong impact on improving the quality of education. The authors discuss the benefits, challenges, and future prospects of incorporating AI in higher education. In Fatema AlDhaen (2022) presents a comprehensive review of the use of AI in higher education, examining various applications and implications of this technology. This includes its use in online teaching and learning, student support services, and data analysis for decision-making. A systematic approach also discusses the potential benefits and challenges of implementing AI in higher education. In Keller et al. (2019) found that machine learning and AI are becoming increasingly prevalent in higher education, with applications ranging from student recruitment and enrollment to learning analytics and personalized learning. In addition, the report highlighted the challenges and opportunities that come with the use of these technologies, such as ethical considerations and the need for specialized training for faculty and staff.

In Hu, Wei (2017) Joseph E. Aoun focused on influence of AI on higher education, received praise for its insightful analysis and thought-provoking ideas. It was well received by both educators and students, sparking important conversations about the role of universities in preparing students for the ever-evolving job market. In Verma et al. (2023) published an article in *International Journal on Recent and Innovation Trends in Comp. & Comm.* on their innovative use of IoT and AI technology for student tracking. Its IoT enabled on time appearance system, which utilized an AI camera and deep learning, aimed to improve student attendance and monitoring in educational institutions. The authors highlighted the potential benefits and challenges of implementing such a system in schools and universities. In Vashishth et al. (2023) delved into the potential of computer vision in accurately recognizing human emotions and provided a comprehensive analysis of various studies on the topic. Authors concluded that vision of computer has immense strengths in accurately detecting and classifying human emotions, making it a valued tool in various fields like psychology, selling, and computer-human interaction. In Cecilia Ka Yuk Chan (2023) proposed a framework for integrating AI policy education into university teaching and learning. The framework aimed to skill students with necessary familiarity and skills to steer compound ethical and policy issues neighboring AI. In Marcinkowski et al. (2020) explore the implications of AI fairness (or lack thereof) on students' reactions and the reputation of educational institutions. Their findings highlight reputation of handling and mitigating potential biases in AI systems to ensure fair and ethical admissions processes. In Grájeda et al. (2024) delved into use of AI tools in higher education and how it impacts students. Through their research, they developed a synthetic index to measure the effectiveness of AI tools in the educational setting, providing valuable insights for educators and policymakers. In Zawacki-Richter et al. (2019) conducted a systematic review of existing research on this topic and found that while there is a growing body of literature on AI in higher education, there is a lack of attention toward the perspective of educators. The authors call for more research to be conducted on the impact of AI on teaching and learning, as well as the role of educators in implementing and utilizing AI in the classroom. In Roy et al. (2022) delved into the potential use of AI-based robots in universities for educational purposes and the factors that may influence their adoption. Through their research, they aimed to provide insights and recommendations for universities considering implementing AI-based robots in their curriculum. The researchers first addressed the growing interest in AI and its potential impact on education. They highlighted the benefits of using AI-based robots in classrooms, such as improved student engagement and modified learning experiences. However, they acknowledged concerns and challenges surrounding acceptance of AI in education, such as the displacement of teachers and ethical considerations.

5.3 AI IN HIGHER STUDIES: A TRANSFORMATIVE SHIFT

Integration of AI in higher study marks a transformative shift that holds power to reshape the entire educational paradigm. Traditional models of education are undergoing a profound evolution as AI technologies infiltrate academic and administrative realms. AI's impact on learning outcomes is monumental, introducing personalized approaches that adapt to individual student needs, revolutionizing pedagogical methods, and fostering enhanced student engagement. This transformative shift extends beyond the classroom, infiltrating administrative processes where routine tasks are automated, allowing for more efficient resource allocation and streamlined operations. The adoption of AI-driven analytics provides educators with unprecedented insights into student performance, enabling timely interventions to address challenges and offer tailored support. This not only enhances academic success but also cultivates a dynamic and responsive educational environment.

Moreover, the integration of AI raises fundamental questions to evolving part of educators and the ethical deliberations surrounding data privacy. As universities embrace this transformative journey, they must navigate the delicate balance between technological innovation and ethical responsibility. This paradigm shift, driven by AI, is reshaping higher education into a more adaptive, inclusive, and technologically advanced landscape. The implications are far-reaching, promising a future where education is not only personalized and efficient but also at the forefront of technological advancements, preparing students for the difficulties of digital age.

5.3.1 Overview of AI Integration in Higher Education

Integration of AI in higher education represents a comprehensive and changing process that is fundamentally reshaping the landscape of learning and institutional management. At its core, AI integration involves leveraging advanced technologies to enhance various facets of the educational experience. Academically, AI is deployed to personalize learning journeys, catering to the diverse needs of students through adaptive curriculum design, intelligent tutoring systems, and data-driven insights into individual progress. Beyond the classroom, AI transforms administrative functions, automating routine tasks, optimizing resource allocation, and providing invaluable analytics for informed decision-making. This holistic approach aims to create an environment that is not only academically enriched but also operationally efficient. However, this transformative integration comes with challenges, including ethical attentions, data privacy concerns, and need for educators to adapt to evolving roles. The overview of AI integration in higher study encompasses a multidimensional exploration of these opportunities and challenges, shedding light on the potential for innovation and improvement while recognizing the imperative of responsible and ethical implementation.

5.3.2 Current Landscape and Adoption Trends

Current landscape of AI adoption in higher study reflects a dynamic intersection of technological advancements and institutional strategies to enhance educational processes. Across academia, institutions are increasingly embracing AI to revolutionize teaching methodologies, personalized learning experiences, and academic support services. Intelligent tutoring systems, chatbots, and virtual assistants are becoming integral components, offering students personalized assistance and immediate feedback. Administrative functions witness a parallel transformation, with AI streamlining admissions, automating routine administrative tasks, and optimizing resource allocation. Moreover, data analytics powered by AI are providing valuable insights into student performance, helping educators tailor interventions and refine pedagogical approaches. The adoption trends underscore a growing recognition of AI's potential to improve learning outcomes, operational efficiency, and overall institutional effectiveness. However, issues like ethical considerations, data privacy, and need for comprehensive faculty training persist. As universities navigate this evolving landscape, the trends indicate a commitment to harnessing the transformative power of AI to create adaptive, efficient, and student-centric higher education environments (Figure 5.2).

FIGURE 5.2 Current Landscape and Adoption in Higher Education AI.

Source: Author created.

5.3.3 Motivation for AI Integration

The motivation for integrating AI in higher study stems from recognition of its potential to revolutionize and optimize various facets of the learning environment. Educational institutions are motivated by a commitment to providing students with a personalized and adaptive learning experience that aligns with their individual needs, preferences, and pace of learning. AI-driven technologies, such as intelligent tutoring systems and adaptive learning platforms, offer the promise of tailored educational journeys, fostering greater engagement and understanding. Moreover, institutions are motivated by the desire to enhance operational efficiency through the automation of administrative tasks, enabling staff to focus on more complex and strategic aspects of education delivery. The predictive analytics capabilities of AI also motivate educators and administrators to proactively identify students at risk, allowing for timely interventions and support. In essence, the motivation for AI integration lies in its potential to elevate the quality of education, improve learning outcomes, and create a more agile and responsive learning ecosystem that prepares students for the challenges of the digital era.

5.4 IMPACT ON LEARNING OUTCOMES

Impact of AI on learning outcomes in higher academics is profound and multifaceted. By introducing adaptive learning technologies, personalized tutoring systems, and intelligent content delivery, AI tailors educational experiences to individual student needs. This customization, grounded in data-driven insights, facilitates a more engaging and effective learning process. AI also enables educators to identify specific areas where students may struggle, allowing for targeted interventions and support. Moreover, the continuous feedback loop provided by AI fosters a dynamic learning environment where students receive timely assessments and adaptive content, promoting a deeper understanding of subject matter. This personalized approach contributes to improved retention rates, increased student satisfaction, and enhanced academic performance. As AI continues to change, its impact on learning outcomes holds promise of not only individualized education but also the cultivation of critical thinking skills and adaptability, preparing students for a rapidly changing professional landscape.

5.4.1 Personalized Learning Experiences

Personalized learning experiences in higher education, facilitated by artificial intelligence, represent a transformative shift toward tailoring education to unique requirements of every student. AI algorithms analyze distinct learning patterns, assess strengths and slackness, and adapt instructional content accordingly. Intelligent tutoring systems and personalized learning platforms utilize this data to create a dynamic curriculum that aligns with students' learning styles and paces. Through adaptive assessments and real-time feedback, students receive targeted support, fostering a deeper understanding of the material. The result is a student-centric approach that not only enhances comprehension but also promotes self-directed learning and

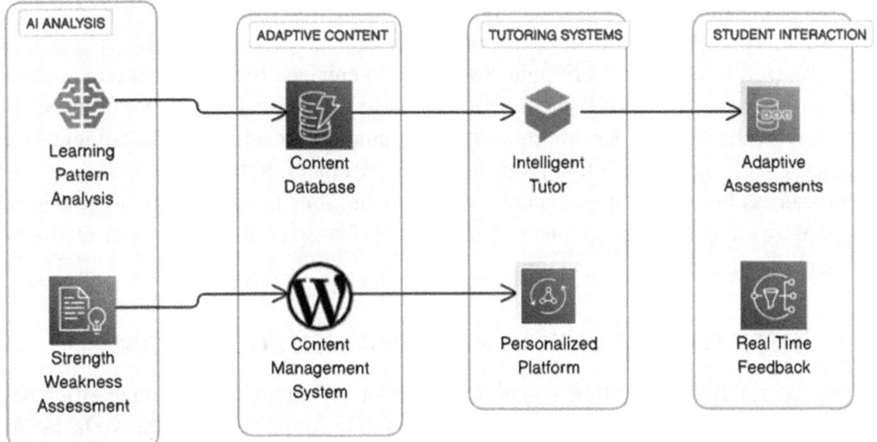

FIGURE 5.3 Personalized Learning Experiences Architecture.

Source: Author created.

critical thinking skills. By using power of AI for personalized learning, higher education institutions aim to create adaptive and inclusive environments that cater to varied needs of their student populations, ultimately fostering a more engaged and empowered community of learners (Figure 5.3).

5.4.2 Enhancing Student Engagement

Enhancing student engagement through the integration of AI in higher education is a pivotal aspect of the transformative impact on learning environments. AI-driven technologies, e.g. chatbots, virtual assistants, and interactive education platforms, contribute to a more dynamic and participatory educational experience. These tools provide immediate responses to student queries, fostering a sense of continuous interaction and support. Additionally, AI-powered analytics offer insights into individual learning patterns, enabling educators to tailor content and interventions to maximize student engagement. Virtual simulations and gamified elements further captivate students' interest, making the learning process more interactive and enjoyable. By leveraging AI to enhance student engagement, higher education institutions aim to create a more vibrant and inclusive learning atmosphere, promoting active participation, collaboration, and a deeper connection between students and the educational content.

5.4.3 AI-Driven Analytics for Student Performance

AI-driven analytics for student performance is revolutionizing the way educational institutions assess and support the academic journey of students in higher education. Artificial intelligence algorithms analyze vast datasets, including academic performance, attendance, and engagement metrics, to derive meaningful insights. These analytics provide educators and administrators with a comprehensive understanding of individual student progress, learning patterns, and potential areas of improvement. By identifying at-risk students early on, AI enables proactive interventions, such as personalized tutoring or additional resources, to enhance their academic outcomes. Moreover, these analytics contribute to data-informed decision-making at the institutional level, allowing for continuous improvement in teaching methodologies and curriculum design. The integration of AI-driven analytics not only optimizes student success but also empowers educational institutions to make more adaptive and responsive education environments tailored to evolving requirements of their diverse student populations.

5.5 OPERATIONAL EFFICIENCY IN HIGHER EDUCATION

Artificial intelligence acting's a pivotal role in enhancing operational efficiency within higher education institutions. By automating routine administrative tasks, AI systems streamline workflows and decrease the burden on administrative staff, allowing them to focus on more complex and strategic aspects of their roles. Chatbots and virtual assistants powered by AI contribute to efficient communication and support services, addressing queries and providing information to students and staff promptly. AI-driven data analytics optimize resource allocation and planning, aiding in budget management and infrastructure optimization. Predictive analytics also enable proactive decision-making, helping institutions anticipate challenges and allocate resources more effectively. The integration of AI in operational processes contributes to a more agile and responsive higher education ecosystem, fostering innovation, and allowing institutions to adapt to evolving demands of digital era.

5.5.1 Automation of Administrative Tasks

The automation of managerial responsibilities in higher education through AI represents a transformative shift in operational efficiency. AI-powered systems streamline and expedite routine administrative functions, reducing the manual workload for administrative staff. Tasks such as data entry, document processing, and scheduling can be automated, allowing staff to allocate their time and efforts to more strategic and complex responsibilities. Chatbots and virtual assistants driven by AI enhance communication and support services, providing quick and accurate responses to inquiries, thereby improving overall efficiency. Additionally, AI algorithms can analyze large datasets to generate insights for better decision-making in areas like resource allocation, budgeting, and strategic planning. The automation of administrative tasks not only increases productivity but also contributes to a more responsive and adaptive administrative framework within higher education institutions.

5.5.2 Resource Allocation and Optimization

Resource allocation and optimization in higher education are significantly enhanced through the application of artificial intelligence. AI-driven algorithms analyze various data sources, including enrolment trends, course demand, and faculty availability, to optimize the allocation of resources such as classrooms, faculty assignments, and budgetary allocations. This data-driven approach enables institutions to efficiently distribute resources based on demand, student needs, and operational requirements.

Predictive analytics, a subset of AI, plays a crucial role in anticipating future resource needs and potential bottlenecks. By identifying patterns and trends, AI models help institutions proactively plan for resource allocation, ensuring a balanced and effective distribution.

Moreover, AI contributes to the optimization of administrative processes, reducing inefficiencies and improving overall operational effectiveness. From scheduling to budget management, AI-driven systems improve decision-creation by giving valuable insights resulting from data analysis.

In essence, the integration of AI in resource allocation and optimization empowers higher education institutions to make informed, strategic decisions that align with their goals and improve the overall efficiency of their operations.

5.5.3 Intelligent Systems for Administrative Processes

Intelligent systems, powered by artificial intelligence, are revolutionizing administrative processes in higher education. These systems bring a level of sophistication to administrative tasks, automating routine processes and enhancing overall efficiency.

Automated Data Processing: Intelligent systems excel at handling large volumes of data, automating tasks such as data entry, document processing, and record-keeping. This not only reduces manual effort but also minimizes the risk of errors.

Chatbots and Virtual Assistants: This streamlines communication and support services. They provide quick responses to inquiries, guide users through processes, and offer information, improving accessibility and responsiveness in administrative interactions (Figure 5.4).

Predictive Analytics: By analyzing historical data, predictive analytics models within intelligent systems assist in forecasting future trends. This capability aids in proactive decision-making, allowing institutions to anticipate needs, plan resources, and optimize administrative processes.

Workflow Optimization: Intelligent systems optimize workflows by identifying inefficiencies and proposing streamlined processes. This enhances the overall effectiveness of administrative operations, reducing redundancy and improving the allocation of resources.

Decision Support Systems: AI-driven decision support systems analyze complex datasets to provide administrators with valuable insights. This aids in strategic decision-making related to budget management, resource allocation, and overall institutional planning.

Intelligent systems for administrative processes not only increase efficiency but also contribute to a more agile, data-driven, and adaptive administrative framework in higher education institutions.

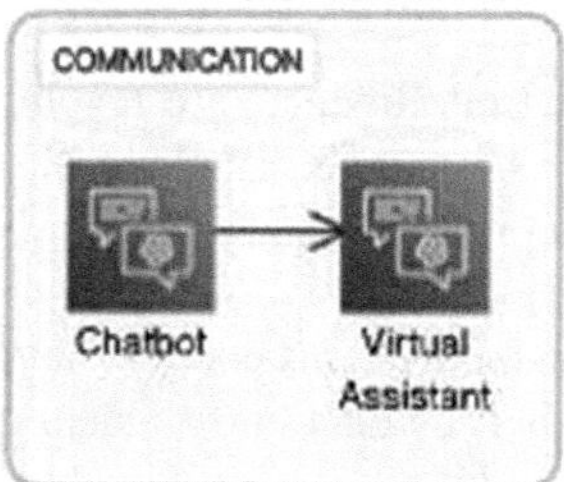

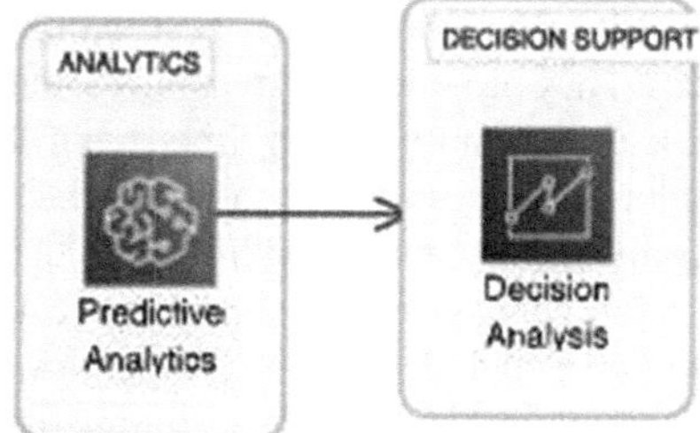

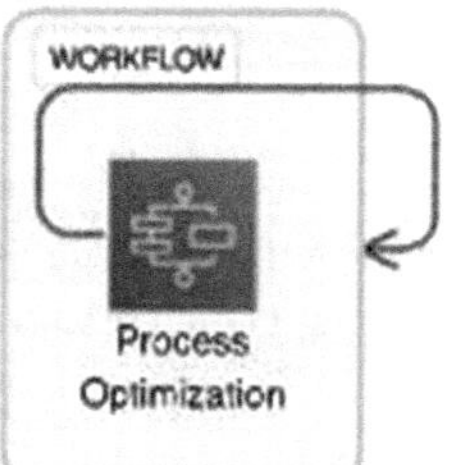

FIGURE 5.4 Intelligent Systems for Administrative Processes.

Source: Author created.

5.6 CHALLENGES AND CONSIDERATIONS

Integration of AI in higher education, while hopeful transformative benefits, comes with a set of issues and considerations that institutions must carefully navigate:

Ethical Concerns: AI applications often involve handling sensitive student data, raising ethical considerations regarding privacy, consent, and data security. Institutions must establish robust ethical frameworks and ensure compliance with relevant regulations to protect rights of students.

Bias and Fairness: AI algorithm can inadvertently perpetuate fakes represent in historical facts, leading to unfair outcomes. Ensuring fairness in AI decision-making processes is a critical challenge, requiring ongoing monitoring, transparency, and efforts to address biases within algorithms.

Data Quality and Accessibility: AI systems heavily depend on quality and accessibility of data. Incomplete or biased datasets can compromise the effectiveness of AI applications. Ensuring data accuracy, relevance, and inclusivity is crucial for the success of AI initiatives.

Staff Training and Preparedness: The successful integration of AI requires a workforce that is knowledgeable and prepared to collaborate with intelligent systems. Providing adequate training for staff and faculty to understand and leverage AI tools is essential for seamless implementation.

Financial Investment: Implementing AI technologies often requires a significant financial investment. Institutions must carefully assess the cost benefit ratio, considering factors such as infrastructure, software development, and ongoing maintenance costs.

Change Management: Introducing AI-driven processes may encounter resistance from staff or faculty accustomed to old-style methods. Actual change management plans are vital to address concerns, communicate benefits, and ensure a smooth transition.

Regulatory Compliance: Higher education institutions must navigate a complex landscape of regulations, including those related to data protection and privacy. Ensuring compliance with relevant laws and standards is crucial to avoid legal complications.

Lack of Standardization: The absence of standardized frameworks for AI in higher education can lead to challenges in interoperability and collaboration between institutions. Establishing industry standards can facilitate a more cohesive and interconnected AI ecosystem.

Student and Stakeholder Engagement: Involving students and relevant stakeholders in AI integration process is crucial for success. Ensuring transparency, addressing concerns, and fostering understanding among all stakeholders contribute to the responsible implementation of AI.

Scalability: Institutions must consider the scalability of AI solutions to accommodate growth in student numbers and evolving needs. Scalability ensures that AI applications remain effective and relevant in the long term.

Navigating these issues requires a holistic method which combines technological expertise, ethical considerations, and an assurance to ongoing improvement. Institutions embracing AI in higher education must prioritize responsible and transparent practices to harness the full potential of these technologies.

5.6.1 Ethical Considerations in AI Adoption

Adoption of AI in various domains raises profound ethical considerations that must be carefully addressed to ensure responsible and fair use of technologies (Figure 5.5).

In context of artificial intelligent adoption, several ethical considerations emerge:

Bias and Fairness: Artificial intelligent algorithms may accidentally perpetuate biases present in historical data, leading to discriminatory outcomes. Ethical AI adoption requires efforts to identify, mitigate, and rectify biases, ensuring fair and equitable treatment across diverse demographic groups.

Privacy Concerns: AI often involves the processing of vast amounts of personal data. Ensuring robust privacy measures, informed consent, and transparent data practices are essential to protect individuals' privacy rights and maintain public trust.

Transparency and Explainability: Ethical AI adoption necessitates transparency in how AI decisions are made. Ensuring that AI algorithms are

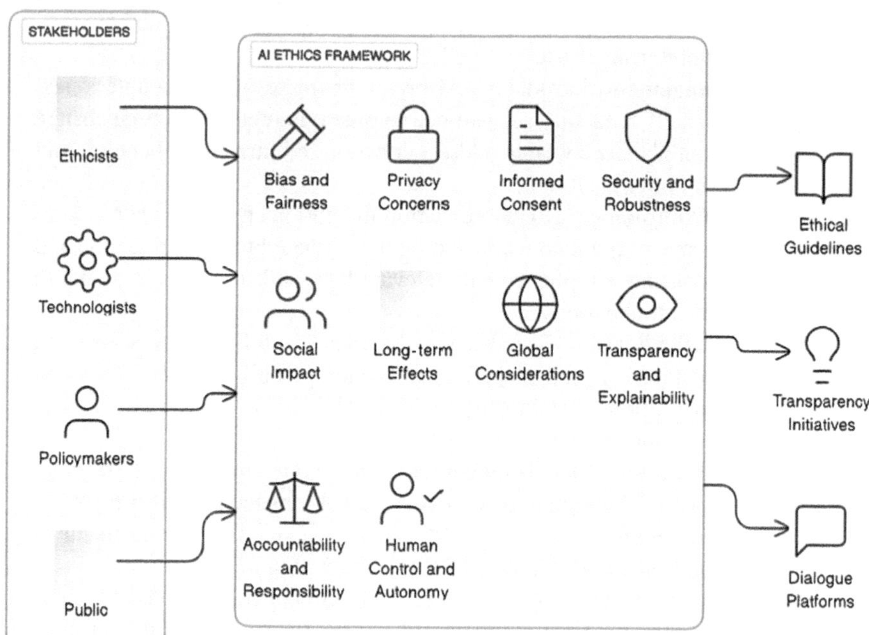

FIGURE 5.5 Ethical Considerations in AI Adoption.

Source: Author created.

explainable and understandable to non-experts promotes accountability and trust in the technology.

Informed Consent: Individuals impacted by AI systems, such as users or subjects of AI applications, should be informed about how their data is used and the implications of AI-driven decisions. Obtaining informed consent is a fundamental ethical principle in AI adoption.

Accountability and Responsibility: Establishing clear lines of accountability for AI systems is crucial. Organizations adopting AI must take responsibility for the outcomes of AI applications, acknowledging and rectifying any negative impacts.

Security and Robustness: Ensuring the security and robustness of AI systems is an ethical imperative. Protecting against adversarial attacks, ensuring data integrity, and preventing unauthorized access are vital considerations in AI adoption.

Social Impact: AI can have wide-reaching societal impacts. Ethical considerations include evaluating how AI adoption may affect employment, social dynamics, and access to resources, and taking steps to mitigate negative consequences.

Human Control and Autonomy: Ethical AI adoption involves preserving human control and decision-making autonomy. AI systems should complement human judgment rather than replace it, and there should be mechanisms for human intervention in critical decisions.

Long-Term Effects: Ethical AI adoption requires organizations to consider the long effects of AI on society. Anticipating and mitigating potential negative consequences ensures responsible and sustainable AI use.

Global Considerations: Organizations adopting AI on a global scale must consider cultural, legal, and ethical differences across regions. Adopting a global perspective ensures that AI applications align with diverse ethical norms and values.

Establishing ethical guidelines, fostering transparency, and promoting ongoing dialogue are essential for shaping a responsible and ethical landscape for AI adoption.

5.6.2 Data Privacy Concerns

The widespread adoption of AI and advanced data analytics has prompted significant concerns regarding data privacy. As organizations harness vast amounts of personal information to train AI models and derive insights, the following data privacy concerns come to the forefront:

Informed Consent: Many AI applications involve the collection and processing of personal data. Ensuring that individuals provide informed consent for the use of their data is crucial. Transparency about how data will be used, the purpose of data processing, and any potential risks is essential for ethical AI practices.

Data Security: Security of personal data is dominant. Data breaches can have severe consequences, leading to unauthorized access, identity theft, or other malicious activities. Organizations must implement robust security measures to safeguard the confidentiality and integrity of the data they handle.

Data Minimization: Adopting principle of data minimization involves collecting only necessary information for a specific purpose. Restraining the scope of data collection helps mitigate privacy risks by reducing the amount of sensitive information at risk.

User Control: Individuals must control over their personal information. This includes the ability to access, correct, or delete their information. Providing users with options to manage their data preferences and offering transparency about data handling practices enhances privacy protection.

Algorithmic Transparency: The opacity of some AI algorithms raises concerns about how decisions are made. Ensuring algorithmic transparency, where the logic and decision-making processes are understandable, helps build trust and allows individuals to comprehend how their data influences outcomes.

Profiling and Automated Decision-Making: AI systems often engage in profiling to make predictions or decisions about individuals. Concerns arise when these profiles lead to automated decision-making without sufficient human intervention. Establishing safeguards to prevent unfair or discriminatory outcomes is essential.

Cross-Border Data Transfers: In a globalized digital landscape, data often crosses international borders. Divergent privacy regulations and standards across regions can create challenges. Organizations must navigate these complexities, adhering to applicable regulations and ensuring data protection during cross-border transfers.

Sensitive Data Handling: The processing of sensitive information, such as health records or biometric data, demands heightened privacy measures. Organizations must implement stringent safeguards and comply with specific regulations governing the handling of sensitive data.

Data Ownership: Clarity about data ownership is crucial. Individuals should understand who owns the data they generate or share with AI systems. Establishing transparent data ownership frameworks helps manage expectations and rights related to personal information.

Regulatory Compliance: Binding to data protection regulations, such as General Data Protection Regulation (GDPR) or similar laws, are non-negotiable. Organizations must stay informed about evolving privacy regulations and adjust their practices to remain compliant.

Addressing these data privacy concerns requires a comprehensive approach encompassing technical, legal, and ethical considerations. Organizations committed to preserving data privacy prioritize transparency, user empowerment, and responsible data stewardship to build and maintain trust in the AI-driven era.

5.6.3 Evolving Role of Educators

Integration of AI in education brings about a transformative shift in the traditional role of educators. As AI methods become increasingly prevalent in educational settings, educators find themselves navigating new responsibilities and opportunities:

Facilitators of Personalized Learning: AI facilitates personalized studying experiences by adapting content, pace, and assessments to discrete student requirements. Educators take on the role of facilitators, leveraging AI-driven tools to tailor instruction, provide targeted support, and address diverse learning styles.

Curriculum Design and Customization: With AI analyzing vast datasets to identify learning gaps and trends, educators contribute to curriculum design by incorporating insights from AI-generated analytics. They collaborate with AI systems to create customized learning pathways that align with both curriculum standards and individual student requirements.

Monitoring and Intervention: AI-driven analytics offer real-time insights into student performance. Educators become proactive monitors, using AI-generated data to identify struggling students, recognize patterns of engagement, and intervene promptly to provide additional support or resources.

Skill Development for AI Literacy: As AI becomes integral to education, educators play a crucial role in fostering AI literacy among students. They guide students in understanding AI principles, ethical considerations, and the implications of AI technologies. Educators facilitate critical thinking about AI, preparing students for a technology-driven future.

Adaptive Assessment Implementation: AI-powered adaptive assessments provide a dynamic evaluation of student progress. Educators collaborate with AI systems to design assessments that not only gauge understanding but also adapt in real-time to challenge and support each student appropriately.

Tech Integration Specialists: Educators become adept at integrating AI technologies into the learning environment. They act as technology integration specialists, selecting and implementing AI tools that enhance instruction, promote engagement, and streamline administrative tasks.

Social and Emotional Learning (SEL) Advocates: While AI addresses cognitive aspects of learning, educators focus on the social and emotional well-being of students. They emphasize the development of interpersonal skills, empathy, and resilience, recognizing the complementary role of human connection in education.

Ethical AI Guides: Educators show a pivotal role in guiding students on ethical use of AI. They foster discussions on responsible AI practices, address biases, and encourage students to consider the societal impact of AI technologies. Educators contribute to shaping ethical considerations within the AI-driven educational landscape.

Continuous Professional Development: Embracing AI in education requires ongoing professional development for educators. They engage in continuous learning to stay abreast of AI advancements, adapt teaching methodologies, and explore innovative ways to leverage AI for improved learning outcomes.

Collaborators with AI Systems: Educators collaborate with AI systems as partners in the educational process. They provide valuable human insights, context, and emotional intelligence that AI may lack. This collaboration enhances the overall effectiveness of educational strategies.

As the role of educators evolves in the era of AI in education, a harmonious partnership between human educators and intelligent technologies emerges. This partnership aims to maximize the benefits of AI while preserving the essential human elements of education, creating a dynamic and adaptive learning environment.

5.7 FUTURE SCOPE

The future scope for AI in higher education is marked by continuous evolution and transformative possibilities. AI is poised to play an increasingly integral role in shaping the educational landscape, with advancements in ML, NLP, and data analytics driving innovation. Personalized learning pathways, adaptive assessments, and AI-driven tutoring systems will become more sophisticated, tailoring educational experiences to individual student needs. The integration of AI into administrative processes will continue to enhance operational efficiency, allowing institutions to allocate resources strategically. Collaborative efforts between human educators and AI systems will redefine pedagogical approaches, emphasizing a harmonious partnership that leverages the strengths of both. As AI technologies mature, ethical considerations and data privacy will remain crucial focal points, necessitating the development of robust frameworks and guidelines. The future of AI holds the promise of a dynamic and learning environment, where technology empowers educators, engages students, and contributes to the continuous improvement of educational outcomes. The ongoing synergy between human intelligence and artificial intelligence will pave the way for a more adaptive, responsive, and learner-centric educational paradigm.

5.7.1 AI's Potential in Shaping Higher Education

Potential of AI in shaping higher education is revolutionary, ushering in a transformative era that transcends traditional paradigms. AI offers a dynamic toolkit that optimizes educational practices, from personalized learning experiences to streamlined processes of administration. By leveraging AI-driven analytics, institutions gain unprecedented insights into student performance, enabling tailored interventions and improved learning outcomes. The infusion of AI in educational technology enhances student engagement, providing adaptive content and individualized support. Moreover, automation of administrative tasks, guided by intelligent systems, results in operational efficiency and resource optimization. AI's role extends beyond mere automation; it acts as a catalyst for redefining the educator's role, fostering collaboration and innovative teaching methodologies. As the educational landscape evolves, the power of AI in higher study presents not only in ornamental efficiency but also in cultivating a more inclusive, adaptive, and student-centric learning environment. The harmonious integration of AI with human expertise promises a future where education is personalized, accessible, and responsive to the diverse needs of learners.

5.7.2 Innovative Applications and Use Cases

Innovative applications and use cases of artificial intelligence in higher education span a diverse range of areas, revolutionizing traditional practices and opening up new possibilities for both educators and students. Adaptive Learning Platforms: AI-powered adaptive learning platforms analyze student performance data to dynamically adjust content, pace, and difficulty levels, providing a personalized learning experience. Virtual Teaching Assistants: AI-driven virtual teaching assistants offer support to students by answering queries, providing additional resources, and offering real-time feedback on assignments (Figure 5.6).

Predictive Analytics for Student Success: Predictive analytics utilize AI algorithms to forecast student success, identifying potential challenges early on and allowing educators to intervene and offer targeted support. Automated Grading Systems: AI-driven grading systems streamline the assessment process, providing quick and consistent feedback on assignments, quizzes, and exams. Intelligent Teaching Organizations: Such arrangements leverage AI to offer tailored tutoring, adapting to individual learning styles and providing targeted assistance. Smart Campus Solutions: AI applications contribute to the creation of smart campuses, optimizing resource allocation, energy management, and security through data-driven insights. Language Processing for Academic Writing: AI-driven language processing tools assist students in improving their academic writing skills, providing grammar checks, style suggestions, and plagiarism detection. Virtual Laboratories: AI enhances virtual laboratory

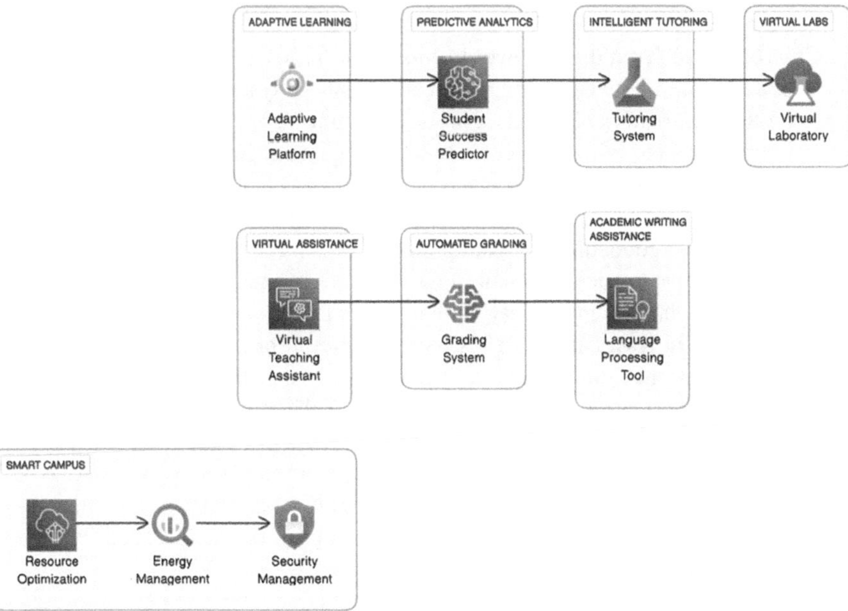

FIGURE 5.6 Innovative AI Applications in Higher Education.

Source: Author created.

experiences, allowing students to conduct experiments in simulated environments, fostering hands-on learning in subjects like science and engineering. These innovative applications showcase how AI is reshaping higher education, fostering efficiency, personalization, and accessibility in the learning journey.

5.7.3 Anticipated Developments in the Field

Anticipated developments in the field of AI in higher education point toward a future characterized by continued innovation, refinement, and widespread integration. Some key anticipated developments include:

Enhanced Personalization: AI will evolve to provide even more personalized learning experiences, tailoring educational content and pathways to individual student needs, preferences, and learning styles. This increased personalization aims to optimize engagement and improve overall learning outcomes.

Advanced Learning Analytics: The field is expected to see advancements in learning analytics, with AI-driven tools becoming more sophisticated in predicting student performance, identifying at-risk students, and offering actionable insights for educators to enhance their teaching strategies.

Ethical AI Practices: As AI becomes more deeply integrated into educational processes, there will likely be a heightened focus on developing and implementing ethical AI practices. This includes addressing biases in algorithms, ensuring transparency in decision-making processes, and safeguarding student data privacy.

Collaborative Human-AI Partnerships: The future may witness a seamless collaboration between human educators and AI systems, fostering a symbiotic relationship where AI assists in routine tasks, grading, and administrative duties, allowing educators to emphasize more on personalized instruction and mentorship.

Expanding Access to Education: AI technologies have the potential to democratize education by making learning resources more accessible. This could involve the development of AI-driven tools that cater to diverse learning needs and provide education in remote or underserved areas.

Innovations in Assessment: AI is likely to play a pivotal role in redefining assessment methods. Adaptive assessments, real-time feedback mechanisms, and AI-powered evaluation tools may become standard, offering a more comprehensive and nuanced understanding of student performance.

Continued Integration of Virtual and Augmented Reality: VR and AR, coupled with AI, may create immersive learning experiences, allowing students to engage with complex concepts in interactive and realistic virtual environments.

Cross-disciplinary Applications: AI in higher education is expected to transcend traditional disciplinary boundaries. Applications could span not only educational technology but also research, administration, and decision-making processes across various academic disciplines.

These anticipated developments underline transformative power of AI in higher education, promising a future where technology augments and enhances the learning experience for students and educators alike.

5.8 CONCLUSION

In conclusion, incorporation of AI in higher education characterizes a paradigm shift with far-reaching consequences for the future of learning and institutional management. The multifaceted impact of AI on personalized learning, operational efficiency, and overall educational outcomes underscores its transformative potential. As we navigate this technological frontier, it becomes evident that AI is not merely a tool but a catalyst for innovation, offering adaptive and tailored learning experiences that cater to diverse student needs. The advent of intelligent analytics allows educators unprecedented insights into student performance, enabling timely interventions and targeted support mechanisms. Despite these advancements, challenges persist, particularly in the realms of ethics, data privacy, and the evolving role of educators. Ethical considerations demand careful scrutiny to mitigate biases and ensure fairness, while safeguarding data privacy remains paramount in an era of increasing digitalization. Moreover, the changing landscape of educators' roles requires ongoing professional development to leverage AI effectively as an aid rather than a replacement. As we look to the future, AI's potential in shaping higher education is not just about efficiency gains but about fostering a dynamic, inclusive, and adaptive learning environment. The envisioned future sees AI not as a disruptor but as an enabler, collaboratively working with educators to empower students and enhance the educational journey. The path forward necessitates a strategic and ethical approach, fostering collaboration among stakeholders to address challenges, embrace innovations, and ensure responsible addition of AI into the fabric of academics. Ultimately, the symbiotic relationship between human intelligence and artificial intelligence holds the promise of a more agile, responsive, and equitable higher education landscape, poised to meet the evolving needs of students and prepare them for the challenges of a rapidly changing world. Through thoughtful implementation and continuous refinement, AI in higher education stands as a beacon guiding the sector toward a future where innovation, inclusivity, and excellence converge in the pursuit of knowledge and educational advancement.

REFERENCES

AlDhaen, Fatema. "The use of artificial intelligence in higher education–systematic review." *COVID-19 challenges to university information technology governance* (2022): 269–285. https://doi.org/10.1007/978-3-031-13351-0_13

Chan, Cecilia Ka Yuk. "A comprehensive AI policy education framework for university teaching and learning." *International Journal of Educational Technology in Higher Education* 20, no. 1 (2023): 38. https://arxiv.org/abs/2305.00280

George, Babu, and Ontario Wooden. "Managing the strategic transformation of higher education through artificial intelligence." *Administrative Sciences* 13, no. 9 (2023): 196. https://doi.org/10.3390/admsci13090196

Grájeda, Alberto, Johnny Burgos, Pamela Córdova, and Alberto Sanjinés. "Assessing student-perceived impact of using artificial intelligence tools: Construction of a synthetic index of application in higher education." *Cogent Education* 11, no. 1 (2024): 2287917.

Hannan, Erin, and Shuguang Liu. "AI: New source of competitiveness in higher education." *Competitiveness Review: An International Business Journal* 33, no. 2 (2023): 265–279. https://doi.org/10.1108/cr-03-2021-0045

Hu, Wei. *Joseph E. Aoun: Robot-proof: Higher education in the age of artificial intelligence.* Cambridge, MA: MIT Press, 2017, 216 pp., £ 19.95 (hbk), ISBN 978-026-20-3728-0." (2019): 1143–1145. https://doi.org/10.1007/s10734-019-00387-3

Keller, Birte, Janine Baleis, Christopher Starke, and Frank Marcinkowski. "Machine learning and artificial intelligence in higher education: A state-of-the-art report on the German University landscape." *Heinrich-Heine-Universität Düsseldorf* (Working Paper, erscheint bei Springer Verlag, Berlin) (2019): 1–31.

Kuleto, Valentin, Milena Ilić, Mihail Dumangiu, Marko Ranković, Oliva MD Martins, Dan Păun, and Larisa Mihoreanu. "Exploring opportunities and challenges of artificial intelligence and machine learning in higher education institutions." *Sustainability* 13, no. 18 (2021): 10424. https://doi.org/10.3390/su131810424

Marcinkowski, Frank, Kimon Kieslich, Christopher Starke, and Marco Lünich. "Implications of AI (un-) fairness in higher education admissions: The effects of perceived AI (un-) fairness on exit, voice and organizational reputation." In *Proceedings of the 2020 Conference on Fairness, Accountability, and Transparency*, pp. 122–130. 2020. https://doi.org/10.1145/3351095.3372867

Muhie, Yimer Amedie, and Abeselom Befekadu Woldie. "Integration of artificial intelligence technologies in teaching and learning in higher education." *Science and Technology* 10, no. 1 (2020): 1–7. https://doi.org/10.5923/j.scit.202001001.01

Pedró, Francesc. "Applications of artificial intelligence to higher education: Possibilities, evidence, and challenges." *IUL Research* 1, no. 1 (2020): 61–76. https://doi.org/10.57568/iulres.v1i1.43

Roy, Rita, Mohammad Dawood Babakerkhell, Subhodeep Mukherjee, Debajyoti Pal, and Suree Funilkul. "Evaluating the intention for the adoption of artificial intelligence-based robots in the university to educate the students." *IEEE Access* 10 (2022): 125666–125678. https://doi.org/10.1109//access.2022.3225555

Singh, Satya Vir, and Kamal Kant Hiran. "The impact of AI on teaching and learning in higher education technology." *Journal of Higher Education Theory & Practice* 12, no. 13 (2022).

Vashishth, Tarun Kumar, Vikas, Bhupendra Kumar, Rajneesh Panwar, Sunil Kumar, and Sachin Chaudhary. "Exploring the role of computer vision in human emotion recognition: A systematic review and meta-analysis." In *2023 Second International Conference on Augmented Intelligence and Sustainable Systems (ICAISS).* IEEE, 2023. https://doi.org/10.1109/icaiss58487.2023.10250614

Verma, Pushpendra Kumar, Vikas Sharma, Prashant Kumar, Shashank Sharma, Sachin Chaudhary, and Preety Preety. 'IoT enabled real time appearance system using AI camera and deep learning for student tracking'. *International Journal on Recent and Innovation Trends in Computing and Communication* 11, no. 6 (10 June 2023): 249–254. https://doi.org/10.17762/ijritcc.v11i6s.6885

Zawacki-Richter, Olaf, Victoria I. Marín, Melissa Bond, and Franziska Gouverneur. "Systematic review of research on artificial intelligence applications in higher education–where are the educators?" *International Journal of Educational Technology in Higher Education* 16, no. 1 (2019): 1–27. https://doi.org/10.1186/s41239-019-0171-0

6 AI in Academic Research
Advances, Opportunities, and Challenges

Almula Umay Karamanlıoğlu
Başkent University, Turkey

6.1 INTRODUCTION

Today, the 21st century is often referred to as the age of technology (Raja & Nagasubramani, 2018). Occupations with intricate responsibilities encompass a blend of structured and unstructured decision-making (Dede, 1988). This leads people to the necessity of utilizing technology. As individuals' tendency to use technology increases, developments in the field of modern technology also manifest themselves. Hence, advances in contemporary technology bring vital developments such as Big Data, cloud computing, machine learning, and artificial neural networks to the agenda. Big Data, along with the rise of cloud computing, artificial neural networks, and machine learning, has led to the emergence of a technology that can simulate human intelligence (Zhai et al., 2021). Along these lines, artificial intelligence (AI) is defined as a simulation of human intelligence processes carried out by computer systems (Gillath, Ai, Branicky, Keshmiri, Davison, & Spaulding, 2020). Precisely, it can be thought as the examination of the processes underlying intelligent behavior (Dede, 1988). It offers systematic capabilities of reasoning based on inputs and learning via the differences of expected outcomes as it predicts and adapts to changes in its ecosystems and stimulus that the system receives from its external environment (Dwivedi et al., 2023:4).

When combined with other technologies, artificial intelligence contributes to numerous sectors such as health, transportation, business, and many other fields (Siau and Wang, 2020). One of these is academic studies in the education sector. Artificial intelligence has impacted academic research in many ways and has also enabled its progress. Artificial intelligence applications have begun to be increasingly used in the education sector, extending beyond the traditional understanding (Chen, Chen, & Lin, 2020a). In educational processes, artificial intelligence technologies are employed in numerous ways and processes regarding automation of diverse tasks and management processes, curriculum development and learning (PK, 1984). In this regard, artificial intelligence in education represents technological advances, theoretical innovations, and pedagogically successful outcomes (Roll & Wylie, 2016).

The future of higher education is shaped by emerging technologies and the computing capacities of smart machines (Popenici & Kerr, 2017). Artificial intelligence

DOI: 10.1201/9781032644509-6

also has the potential to shed light on academic research in numerous fields (M Alshater, 2022). Most of the artificial intelligence technologies used in the education sector have an understanding that reflects current educational assumptions and practices, including the general learning approach (Holmes, Bialik, & Fadel, 2023). The role of technology in higher education is to enhance human thought and the educational process (Popenici & Kerr, 2017). With the increasing use of artificial intelligence technologies, the number of studies published in the academy has increased in this way (Chen, Zou, Xie, Cheng, & Liu, 2022). Machine learning and artificial intelligence continue to emerge increasingly in the academic field (Livberber & Ayvaz, 2023). The increasing use of artificial intelligence technologies in both academia and industry is evident in the conference and participation statistics on this subject (Amini et al., 2020). It also allows issues that are overlooked or hidden with technology to emerge (Holmes, Bialik, & Fadel, 2023). Artificial intelligence comes to the fore with the increasing use of automatic chat robots that provide learning experiences to users and attempts to integrate them into the virtual reference process (Yao, Zhang, & Chen, 2015). Moreover, artificial intelligence enables deep learning and Big Data analysis to accomplish difficult tasks and allow researchers to do more work. Artificial intelligence, which can analyze large amounts of data in research, interpret it and transmit it in features that require many skills such as creating simulations, also affects the efficiency and effectiveness of research (M Alshater, 2022). In this vein, the combination of several factors such as the fact that artificial intelligence is multidisciplinary, brings together technological advances and new solutions makes it attractive for both academics, researchers, and practitioners.

This chapter draws attention to the advancements, opportunities, and challenges that artificial intelligence presents in the academic domain. While the development of this technology brings positive outcomes in academia, it also brings negative consequences. Notably important in the academic field, research and development studies may be disrupted or eroded, and adverse consequences may arise such as making the role of key actors less essential in this sense. This book chapter is a valuable resource that can shed light ethically and effectively on how researchers can promote their knowledge and experience well to advance artificial intelligence in academic research.

6.2 THE EVOLUTION OF AI IN ACADEMIC RESEARCH

Philosophers have proposed knowledge of what it means to be human by talking about the possible power of intelligent machines (Buchanan, 2005). Alan Turing asked in 1950, "Can machines think?" He may be considered one of the people who brought the concept of artificial intelligence to the fore with an animation game he started to test the question (Turing, 1950). Turing attempted to test whether he could think like a human by engaging a written game through a computer program. If the computer or machine can successfully convince a human in the game it is testing, it is considered acceptable accordingly. The term "artificial intelligence" began to gain popularity at the Dartmouth conference in 1956, which was called the artificial intelligence era (Delipetrev, Tsinaraki, & Kostic, 2020). Artificial intelligence emerged in 1959 as a field that specifically tried to understand the nature of human intelligence (Mesquita, 2018). In this manner, artificial intelligence studies gained momentum in the 1950s and 1960s. Studies in the field of artificial intelligence began to be considered under the name of symbolic artificial intelligence and knowledge-based systems in the 1980s (Delipetrev, Tsinaraki, & Kostic, 2020). Since researchers in this field have tried to model human intelligence through logical concepts and symbols. Therefore, logic-based systems and decision-making systems have begun to be developed. Artificial intelligence researchers began to develop and use more complex models between 1990 and 2010 (Delipetrev, Tsinaraki, & Kostic, 2020). The importance of the concepts of machine learning and deep learning which have gained prominence in these years has come to the fore accordingly. Multilayer neural networks have begun to be studied on more complex data structures and have led to the development of data-based learning models using more complex datasets. While machine learning is considered as an approach that paves the way for the emergence of future data and behaviors by making computers and machines learn from past data or behaviors, the concept of deep learning is considered a branch of machine learning that allows the use of artificial neural network techniques and algorithms for education and training (Alaskar & Saba, 2021).

Advancements in different disciplines and branches of science enable the use and application of artificial intelligence in academic research. In this regard, inventions in many disciplines have been influenced by artificial intelligence. Artificial intelligence is considered a technological science that investigates the theory of human intelligence and its related methods, techniques, and applications to expand and simulate (Ning & Yan, 2010). Artificial intelligence can be defined as machine intelligence as opposed to natural intelligence exhibited by humans (Delipetrev, Tsinaraki, & Kostic, 2020). The term "artificial intelligence" covers computer systems that mimic human abilities and may perform related tasks (Auernhammer, 2020). AI is defined as machine intelligence or intelligence demonstrated by machines, in contrast to the natural intelligence displayed by humans (Delipetrev, Tsinaraki, & Kostic, 2020:5).

Artificial intelligence technologies require computing power, 3D printing, internet of Things, data lakes, cloud computing, and advanced workforce to perform effectively (West & Karsten, 2015). In this manner, managing the resources of artificial intelligence technologies is essential for the advancement of studies in the field of artificial intelligence. Artificial intelligence can learn from historical data and recognize behavior (Alaskar & Saba, 2021). Researchers have pointed out the important roles of artificial intelligence in the discovery, improvement, and evaluation of new methods and processes (Cyranoski, 2019). Artificial intelligence technologies are thought to enable potential researchers to save time by getting rid of mundane tasks in institutions (AJE, 2018). Furthermore, it can be viewed as beneficial in creating an interactive and adaptive learning environment by providing ease of translation. AI algorithms continuously improve through machine learning, minimizing the likelihood of errors that may occur due to human fatigue or oversight (Qodirov, 2023:123). As emphasized before, artificial intelligence technologies are used in numerous aspects of the academic field. It is used in numerous activities such as course material production, performance analysis, conference and seminar creation, scientific research, and so on. It can be viewed that it has an impact on accelerating scientific discoveries such as learning processes, preparing course content, analyzing scientific research, designing models, and interpreting the related results. The ability of AI to understand, interpret, and generate human language has opened new avenues for the enhancement of academic research and education (Pinzolits, 2023:38).

6.3 AI TECHNOLOGIES AND TOOLS OF AI IN VARIOUS ACADEMIC DISCIPLINES

Artificial intelligence consists of the combination of many technologies that enable machines to perceive, act, and learn (Bowen & Morosan, 2018). In recent years, it has been thought to bring significant developments in the field of education, especially with its potential to support learning, as it has an algorithmic-based power in matters such as predictions, suggestions, diagnoses, and decisions regarding artificial intelligence technologies (Hwang, Xie, Wah, & Gašević, 2020). Rapid developments in artificial intelligence technologies have had a profound impact on the fields of politics, science, economics, and education (Luan et al., 2020).

Learning sciences are filled with various theories and methods regarding the development and implementation of educational technology (Kay & Luckin, 2018). The use of artificial intelligence technologies is thought to have a significant increase in educational practices (Chen, Xie, & Hwang, 2020c). Artificial intelligence supports individuals in collaboration such as thinking, feeling, planning, monitoring, and controlling (Cranefield, Winikoff, Chiu, Li, Doyle, & Richter, 2023). It can facilitate the use of language tools required in the academic field and perform various tasks such as text translation and summarization. It helps in completing a certain workflow such as preparing reports and creating automatic workflows. There are numerous tools that incorporate artificial intelligence technologies in academic research. Studies on artificial intelligence systems and chat robots frequently appear in the academic field (Zhang & Aslan, 2021). Owing to the language processing (NLP) techniques included in artificial intelligence technologies, it provides convenience to practitioners in important features such as analysis, understanding, and interpretation of a different language. It can be acknowledged that crucial in the development of written and voice interfaces thanks to technologies that require user interaction such as chatbots.

There are studies that ChatGPT is a publicly available tool developed by OpenAI (Kirmani, 2022). However, certain policies and restrictions may be imposed on the use and development of such applications. ChatGPT is one of the applications that can answer users' questions and perform language processing tasks thanks to its language processing ability. In this vein, ChatGPT is thought to provide users with convenience such as correcting grammatical errors (Stokel-Walker & Van Noorden, 2023). It can be recognized that it is beneficial in academic research in performing many tasks regarding language and in performing complex tasks. ChatGPT has the potential to greatly impact academia, scientific research, and publishing (Lund, Wang, Mannuru, Nie, Shimray, & Wang, 2023).

When integrated into higher education, AI includes a range of technologies and methods in various forms such as machine learning, language processing, data mining, neural networks, and algorithms (Baker, Smith, & Anissa, 2019). It has worldwide importance, notably in the learning industry (Chen, Xie, Zou, & Hwang, 2020b). Research reveals that chatbot technologies can improve learning processes and student interaction (D'Mello, Olney, Williams, & Hays, 2014). Chatbots attract attention as automatic conversational tools that regulate the language processing and machine learning algorithms necessary to communicate with users (Kooli, 2023). Studies show that using chatbots greatly increases students' performance and retention of information (Alotaibi, Al-Shehri, Al-Harbi, & Al-Mutairi, 2020). They can provide instant responses to users, provide a personalized learning environment, monitor users' performance, and provide various reminders for them to review. In this vein, modern chat robots understand people's written and spoken language, improve the answers they give to users during their interactions, and store the information they receive (Pratt, 2017). A new training methodology is proposed to improve the text comprehension ability in GPT-4 (Okuyama & Suzuki, 2023). As emphasized before, studies are showing that ChatGBT increases the potential for advanced learning in higher education (Sullivan, Kelly, & McLaughlan, 2023). In this vein, many people prefer to use artificial intelligence applications because they provide benefits in various subjects such as chatting, answering questions, creating articles, creating content, and so on (Dwivedi et al., 2023).

6.4 FUTURE ADVANCES, OPPORTUNITIES, AND CHALLENGES

6.4.1 Future Advances

Considering that significant advances in machine learning and artificial intelligence are opening up new possibilities and challenges for higher education, it is meaningful to remember that the education sector is a human-centered endeavor rather than a technology-centered solution (Popenici & Kerr, 2017). Artificial intelligence in education provides a more adaptable learning environment by responding to the qualified needs of students and educators in language acquisition and optimizing teaching (Malik, Pratiwi, Andajani, Numertayasa, Suharti, & Darwis, 2023).

In the swiftly evolving academic environment, artificial intelligence applications are gaining more influential for both academics and students. These applications can come to the fore in research fields and play a determining role in the course of the academic world. Artificial intelligence applications manifest themselves in a variety of ways in academia. Some studies show that it can provide real-time feedback, numerous personalized learning paths, gamification, and increased accessibility (Pokrivčáková, 2019). Moreover, it enables students to develop language skills interactively by providing a realistic speaking environment (Malik, Pratiwi, Andajani, Numertayasa, Suharti, & Darwis, 2023). Artificial intelligence facilitates language learning in an inclusive way (Roll & Wylie, 2016). Nonetheless, artificial intelligence applications support students' self-efficacy in the field of writing (Gayed, Carlon, Oriola, & Cross, 2022). Since it can be viewed that it offers several trainings to correct students' mistakes, make their writings more readable, offer suggestions, and improve their writing skills. It also supports users in various areas such as developing creative writing skills and editing texts. In addition, compared to traditional language classes, artificial intelligence-supported language tools provide economic savings to educators and students (Pokrivčáková, 2019). It also helps to reach more users by making language learning processes more accessible and convenient. Moreover, the digital library supports practitioners in providing better training through voice assistants and robots. In this vein, it can reduce additional payments that may occur. Advances in artificial intelligence play an influential role in solving university budget cuts (Popenici & Kerr, 2017). This could lead to increased efficiency in research areas and the ability to manage resources conclusively.

Conversely, advancements in the field of artificial intelligence technology support the idea that it guides students in its correct use by examining the effects of artificial intelligence in detecting plagiarism (Chaudhry, Sarwary, El Refae, & Chabchoub, 2023). Studies have found strong relationships by comparing the human rating system and the artificial intelligence rating system in the artificial intelligence-supported automatic article evaluation system (Stahl et al., 2023). These results show that the evaluation system is supported in terms of its reliability (Malik et al., 2023).

Nonetheless, knowing the social impacts of chatbot utilization is essential in guiding forthcoming studies and design (Følstad et al., 2021). Chatbots are thought to provide significant benefits as one of the artificial intelligence technologies in giving instant feedback to students and providing personalized learning (Kooli, 2023). It can be used to exchange information, use data analysis and reporting processes effectively, collect large datasets, and study human behavior in a variety of experiments and applications. The use of chatbots in the academic field attracts the attention of both academicians and students in many areas. Studies show that using chatbots makes students' first years at university easier and increases their commitment to the course (Studente, Ellis, & Garivaldis, 2020). Notwithstanding, it can be thought that it provides many benefits to practitioners by providing support in solving various questions and using various communication channels. Below are some of the popular artificial intelligence applications used in the academic field.

6.4.1.1 OpenAI

OpenAI is a research laboratory that provides great benefits to artificial intelligence technologies, including its highly developed language model GPT-3 (Lund, Wang, Mannuru, Nie, Shimray, & Wang, 2023: Taecharungroj, 2023). Nevertheless, OpenAI has also proposed ChatGPT, which uses natural language processing, as a chat robot (Lund, Wang, Mannuru, Nie, Shimray, & Wang, 2023). ChatGPT is the latest development in the group of systems known as "chatbots" (Taecharungroj, 2023:1). It is widely used in many fields with its low cost, convenience, and improved user experience (Thorat & Jadhav, 2020).

6.4.1.2 ChatGPT

ChatGPT, an artificial intelligence chatbot, is widely used in education and various academic fields. It is the brainchild of OpenAI, a California-based artificial intelligence company (Stokel-Walker, 2022). These chat robots, which have different versions such as ChatGPT 3 and ChatGPT 4, provide users with convenience in various subjects such as language use, reasoning, creativity, and problem-solving.

6.4.1.3 Scite.ai

Artificial intelligence technologies present supporting and distinguishing solutions to readers in providing a more detailed understanding of the context in which a quote is made (Bakker, Theis-Mahon & Brown, 2023). Citations are a critical component of scientific publishing, linking research findings across time (Nicholson, etc., 2021:882). Citation numbers are often employed to assess the impact of articles (Brody, 2021). Smart Citations reveal how a scientific paper has been cited by providing the context of the citation and a classification system describing whether it provides supporting or contrasting evidence for the cited claim, or if it just mentions it (Nicholson, etc., 2021:883). Nevertheless, it extracts context by identifying quotations within the text of an article and offers a deep learning model to classify the cited article as supporting or arguing (Nicholson et al., 2021). In this vein, the ability to distinguish between citation statement types is important in evaluating the accuracy of research articles (Sterner, 2021).

6.4.1.4 Elicit

Elicit can be considered one of the frequently preferred applications in the academy, especially for the development and learning of language-related skills. It has a variety of different contents and offers a variety of materials. It can be acknowledged that it helps the practitioners who use the application to follow their progress and various skills. At the same time, it can provide the advantage of providing a different experience to people by providing a personalization environment tailored to the goals and needs of practitioners.

6.4.2 Opportunities

Artificial intelligence technologies are consistently expanding their presence in various aspects of society (Vargas-Murillo, de la Asuncion, & de Jesús Guevara-Soto, 2023). The integration of AI systems and chatbots into the academic field has gained significant attention in recent years (Kooli, 2023:1). The development of artificial intelligence technologies has managed to attract people's attention both in education, academia, and various research fields. Lately, ChatGPT, which offers artificial intelligence applications, has managed to attract the attention of various fields, including education (Tlili et al., 2023).

With the increasing popularity of artificial intelligence applications, chatbots have attracted the attention of many sectors including education (Lestari & Wicaksono, 2023). As many academics and students started to use ChatGPT, Bing Chat, Bard, Ernie, and other chatbot tools, they have come under the influence of developments in technology and have managed to influence higher education accordingly (Rudolph, Tan, & Tan, 2023). Chatbots are viewed as a technology that provides and improves learning and teaching experiences (Jeon, 2023). Although there are studies on the positive and negative aspects of chatbots, there are studies that support the positive developments of chatbots in the field of education. It assists practitioners in obtaining learning resources, translation support, grammar, and language structure by providing immediate answers to practitioners. Studies underline that ChatGPT can be cultivated as a second language (Barrot, 2023). In this direction, studies on artificial intelligence show that the use of artificial intelligence in language education provides many benefits (Makeleni, Mutongoza, & Linake, 2023: Nghi, Phuc, & Thang, 2019). Some of these can be considered personalized learning paths, gamification, increased accessibility, and real-time feedback (Pokrivčáková, 2019). Artificial intelligence-supported language tools make it easier for students and practitioners to provide data-based information by optimizing the learning process (Marais, 2021). In this vein, it can recognize repeated patterns in data. Therefore, it can be considered to provide convenience to practitioners in various fields such as text analysis, image, and audio processing. Since artificial intelligence systems are designed to process large amounts of data, learn from patterns, and make predictions from these patterns (Holmes, & Tuomi, 2022).

6.4.3 Challenges

The integration of artificial intelligence into education brings with it some challenges and obstacles by virtue of technological change (Firat, 2023). Despite many advantages in the field of artificial intelligence, there are certain challenges with the utilization of these technologies. Although it is effective in accelerating and developing scientific studies, it also creates dangers that cannot be underestimated in that way (Dupps, 2023). The challenges include developing comprehensive public policies for sustainable development, ensuring inclusion and equity, preparing teachers for AI-powered education, developing quality and inclusive data systems, making research on AI in education significant, and addressing ethical concerns related to data collection, use, and dissemination (Kooli, 2023:3).

There are numerous ethical issues associated with the use of artificial intelligence technologies. These various ethical issues such as plagiarism, data privacy, and security come to the fore. Studies on the challenges of artificial intelligence have underlined that the utilization of artificial intelligence technologies may create outcomes such as plagiarism (Dupps, 2023). One of these causes plagiarism by creating new data using the same or similar content. Artificial intelligence technologies used in academic texts bring with them many ethical problems such as copyright, plagiarism, and authorship problems (Liebrenz et al., 2023). Besides these problems, there are also discussions about data privacy and security (Rodrigues, 2020). However, it also causes various problems such as accountability, prejudices, and ethical and social concerns (Furey & Martin, 2019). As AI solutions have the potential to structurally change university administrative services, the realm of teaching and learning in higher education presents a very different set of challenges (Popenici & Kerr, 2017:2). Integrating artificial intelligence in every higher education institution may not always be quick or easy. Challenges and limitations in this integration process can be considered as relevant in revealing technical problems.

6.4.3.1 Academic Dishonesty

One of the many problems that arise after artificial intelligence systems are frequently used, especially in the academic field, is academic dishonesty. Students tend to cheat on their homework by using some artificial intelligence systems, especially in articles that are not their work (Dehouche, 2021). This increases the likelihood of plagiarism for assessment in higher education (Cotton, Cotton, & Shipway, 2023). Plagiarism results from students' dishonest behavior (Klein, 2011). This may cause various problems such as authorship, originality, ownership of texts, and copyright (Scollon, 1995). Methods used and adopted by artificial intelligence developers are vital accordingly. Artificial intelligence developers may turn to using the same or similar source from another source. Insufficient and unauthorized use of data sources can also be considered as one of the reasons that cause plagiarism. In this vein, the causes of plagiarism cases may be due to different reasons. One of these stems from the lack of knowledge about quoting and quoting (Belter & du Pré, 2009). It can be caused by many factors such as easy access to information, pressure on academic publications, writing articles under stress, and lack of awareness (Mohammed et al., 2015). While students at universities tend to copy a sentence or text of an article, plagiarism reports appearing in scientific publications can present themselves as copying research (Li, 2013). Plagiarism can also create a source of pressure for academics (Sutherland-Smith, 2005). The pressure to publish in the academic field can lead to different problems, such as academics automatically using different sources such as articles or theses. The use of these tools may become more attractive due to various reasons such as intense work pressure, reduced working hours, or easy access to information. There are useful programs used to minimize plagiarism. One of these is the Turnitin program (Ismail & Jabri, 2023). These programs are no longer sufficient with the development of artificial intelligence technologies. Since studies have found that some of the article-writing tools of artificial intelligence systems prevent plagiarism detectors by rearranging sentences and converting them into text (Cotton, Cotton, & Shipway, 2023).

6.4.3.2 Limited Use of Language

Although the artificial intelligence system used in the language field shows good development, peculiarly in English and widely used languages, it seems to make it difficult for many people in the Global South regions to access these services (Liang et al., 2022). Likewise, in the Global South, language-related resources cannot be met due to infrastructure and economic difficulties, and, therefore, it becomes difficult to access quality language learning materials by using artificial intelligence systems with limited language options (Taylor & Kochem, 2022). Applications related to the portability of speech and language technologies for languages with limited resources can be considered as areas where more research is needed to improve language discrimination (Besacier, Barnard, Karpov, & Schultz, 2014). The digital divide not only hinders language learning opportunities, but also does not enable the development of tools used for indigenous languages (Taylor & Kochem, 2022). Conversely, the development of artificial intelligence systems for less spoken languages has begun to increase (Nemorin, Vlachidis, Ayerakwa, & Andriotis, 2023).

6.4.3.3 Prejudice

One of the major problems in artificial intelligence tools is bias. Biases may arise through the training data required to develop artificial intelligence-supported technologies (Luengo-Oroz et al., 2021). Chatbots are trained on data from past user interactions and can continue to learn from the data collected along the way (Atkins, Badrie, & van Otterloo, 2021). Such training reveals problems of giving biased answers based on incorrect data (Atkins, Badrie, & van Otterloo, 2021). Conversely, biases that may cause inequality may arise in the early stages of artificial intelligence application development (Luccioni, Bullock, Pham, Lam, & Luengo-Oroz, 2020). The first of these may arise by affecting the framing of the problem to be addressed by artificial intelligence and the scope of the work, and secondly, biases may arise in the way the data used in the artificial intelligence system is labeled (Luengo-Oroz et al., 2021). In addition, technological limitations and design can be instrumental in shaping prejudices in society and excluding some groups (Feine, Gnewuch, Morana, & Maedche, 2020). Artificial intelligence systems can learn biases from training data, and accordingly, the data may tend to influence the models' decisions. They can exclude groups and make decisions regarding exclusion in places that do not have sufficient technological power. Furthermore, the use of artificial intelligence systems in language proficiency assessments shows that these assessments may cause bias. Studies have drawn attention to the error of non-standard language variations that may occur in artificial intelligence systems (Mphahlele & McKenna, 2019). It may occur due to reasons such as dialect specific to social groups, language differences, and different languages. Misleading and biased results may occur in understanding and detecting non-standard language differences. Conversely, a lack of transparency in the use of these tools and limited supervision may lead to inappropriate use, which may negatively affect learning outcomes (Winke & Isbell, 2017).

6.4.3.4 Lack of Confidence

Individuals, organizations, and society invest in and use information systems for specific reasons such as product service development, effective management, and cost reduction (Li, Valacich, & Hess, 2004). These systems bring their vulnerabilities and risks with the development of technology. In this vein, the relationship between trust and technology has a critical value for both individuals and organizations. Researchers in different disciplines define trust according to their context (McKnight, Choudhury, & Kacmar, 2002). Trust can also be defined as part of how people interact with technology (Li, Hess, & Valacich, 2008). Given the progress in the development of AI, a correspondingly sophisticated understanding of trust in the technology is required (Choung, David, & Ross, 2023:1). Trust is essential for market adoption of new technologies (Corritore et al., 2005).

The use of artificial intelligence and technological tools may cause several problems regarding the reliability and confidentiality of technologies. These may cause problems such as ethical violations, data privacy violations, discrimination, and prejudice that may occur during the use of technologies. Artificial intelligence may have different ethical problems and social impacts depending on the region (Zhu, 2022). Research supports that the data and coding process of some artificial intelligence applications, notably in language models, may contain biases related to gender, race, ethnicity, and disability (Zhao et al., 2019). Gender, race, and ethnicity situations that often arise in language applications may indicate that biases related to different factors may arise during the coding process. Studies conducted in the academic field can contribute to the ethical and reliable progress of the development and implementation of artificial intelligence technologies. It can conduct a detailed examination of academic and ethical issues and easily detect standard or faulty algorithms. These algorithms can even offer suggestions about social inequalities. To reduce trust-related problems, it is essential to take various precautions, be transparent, and maintain reliability. When transparent, fair, and accurate services are guaranteed, users are more likely to perceive higher credibility in the news (Shin, 2021:548).

6.4.3.5 Lack of Accountability and Transparency

Computer systems use a set of algorithms and a set of rules that determine behavior (Weizenbaum, 1976). Accountability regarding artificial intelligence is widely discussed in the literature (Collina, Sayyadi, & Provitera, 2023). Despite the increasing popularity of artificial intelligence, little is known about people's perception of trust and the processes of creating a sense of trust thanks to algorithmic features (Shin, 2020). Personalized algorithms influence what people choose in the past, present, and future (Shin, 2021). Transparency can be used together with concepts such as explainability, interpretability, openness, and accessibility (Felzmann, Fosch Villaronga, Lutz, & Tamò-Larrieux, 2019). Algorithmic transparency is related to terms such as "algorithmic visibility," "explainability," and "interpretability" (Shin, 2021:547). Although transparency sometimes seems to be agreed upon in other disciplines, questions such as what it means and to what extent it is useful can often not be explained regularly (Felzmann, Fosch-Villaronga, Lutz, & Tamò-Larrieux, 2020). Understanding the mechanisms related to artificial intelligence technologies and how decisions are made about them can be considered important in explaining the accountability and transparency of these technologies. Since how data is collected and processed is of great importance to users in managing and disclosing data sources. To develop artificial intelligence applications, transparency, ethics, and privacy issues need to be resolved (Schiff, Rakova, Ayesh, Fanti, & Lennon, 2020).

6.5 ETHICAL AND RESPONSIBLE AI IN ACADEMIA

Computer systems use a set of algorithms and a set of rules that determine behavior (Weizenbaum, 1976). Accountability regarding artificial intelligence is widely discussed in the literature (Collina, Sayyadi, & Provitera, 2023). Despite the increasing popularity of artificial intelligence, little is known about people's perception of trust and the processes of creating a sense of trust thanks to algorithmic features (Shin, 2020). Personalized algorithms influence what people choose in the past, present, and future (Shin, 2021). Transparency can be used together with concepts such as explainability, interpretability, openness, and accessibility (Felzmann, Fosch Villaronga, Lutz, & Tamò-Larrieux, 2019). Algorithmic transparency is related to terms such as "algorithmic visibility," "explainability," and "interpretability" (Shin, 2021:547). Although transparency sometimes seems to be agreed upon in other disciplines, questions such as what it means and to what extent it is useful can often not be explained regularly (Felzmann, Fosch-Villaronga, Lutz, & Tamò-Larrieux, 2020). Understanding the mechanisms related to artificial intelligence technologies and how decisions are made about them can be considered important in explaining the accountability and transparency of these technologies. Since how data is collected and processed is of great importance to users in managing and disclosing data sources. To develop artificial intelligence applications, transparency, ethics, and privacy issues need to be resolved (Schiff, Rakova, Ayesh, Fanti, & Lennon, 2020).

6.6 CONCLUSION

The challenges and threats that arise with the growing prevalence of artificial intelligence have an impact on the academic field. Although advancements in the field of artificial intelligence presents new opportunities in academic studies, they also bring challenges and constraints. Advances and developments in the field of artificial intelligence have enabled to error-free utilization of data, enhancing the data set to be more efficient. In this manner, the quantity and quality of scientific studies have increased and provided the opportunity to reach more people with expanded datasets.

Alongside the positive developments in the academic field, there are also adverse consequences and threats. These adverse consequences and threats may lead to a decrease in scientific reliability and the emergence of some ethical problems. The academic field, research and development studies may be disrupted or eroded, and adverse consequences such as diminishing the influence of key actors. Conversely, the rising prevalence of artificial intelligence applications brings to the agenda the idea that some academic roles may undergo transformation and be replaced by artificial intelligence tools. There may be a possibility that academic values in the scientific sense will decrease and may have a multifaceted impact on the academic field. In this manner, it is essential to identify the advances, opportunities, and threats in the field of artificial intelligence when assessing both the positive and adverse consequences.

REFERENCES

AJE (2018). *Peer review: How we found 15 million hours of lost time.* https://www.aje.com/arc/peer-review-process-15-million-hours-lost-time/

Alaskar, H., & Saba, T. (2021). Machine learning and deep learning: A comparative review. *Proceedings of Integrated Intelligence Enable Networks and Computing: IIENC* 2020, 143–150.

Alotaibi, R., Al-Shehri, S., Al-Harbi, R., & Al-Mutairi, M. (2020). Enhancing learning outcomes through chatbot technology in computer science education. *Education and Information Technologies*, 25(6), 5167–5183.

Alshater, M. (2022). Exploring the role of artificial intelligence in enhancing academic performance: A case study of ChatGPT. *Available at SSRN.*

Amini, L., Chen, C. H., Cox, D., Oliva, A., & Torralba, A. (2020). Experiences and insights for collaborative industry-academic research in artificial intelligence. *AI Magazine*, 41(1), 70–81.

Atkins, S., Badrie, I., & van Otterloo, S. (2021). Applying Ethical AI Frameworks in practice: Evaluating conversational AI chatbot solutions. *Computers and Society Research Journal*, 1.

Auernhammer, J. (2020). Human-centered AI: The role of human-centered design research in the development of AI.

Baker, T., Smith, L., & Anissa, N. (2019). *Educ-AI-tion rebooted. Exploring the future of artifcial intelligence in schools and colleges*. Retrieved from Nesta Foundation website: https://media.nesta.org.uk/documents/Future_of_AI_and_education_v5_WEB.pd

Bakker, C., Theis-Mahon, N., & Brown, S. J. (2023). Evaluating the accuracy of scite, a smart citation index. *Hypothesis: Research Journal for Health Information Professionals*, 35(2), 1–9.

Barrot, J. S. (2023). Using ChatGPT for second language writing: Pitfalls and potentials. *Assessing Writing*, 57, 100745.

Belter, R. W., & du Pré, A. (2009). A strategy to reduce plagiarism in an undergraduate course. *Teaching of Psychology*, 36(4), 257–261.

Besacier, L., Barnard, E., Karpov, A., & Schultz, T. (2014). Automatic speech recognition for under-resourced languages: A survey. *Speech Communication*, 56, 85–100.

Bowen, J., & Morosan, C. (2018). Beware hospitality industry: The robots are coming. *Worldwide Hospitality and Tourism Themes*, 10(6), 726–733.

Brody, S. (2021). Scite. *Journal of the Medical Library Association: JMLA*, 109(4), 707.

Buchanan, B. G. (2005). A (very) brief history of artificial intelligence. *Ai Magazine*, 26(4), 53–53.

Chaudhry, I. S., Sarwary, S. A. M., El Refae, G. A., & Chabchoub, H. (2023). Time to revisit existing student's performance evaluation approach in higher education sector in a new era of ChatGPT—A case study. *Cogent Education*, 10(1), 2210461.

Chen, L., Chen, P., & Lin, Z. (2020a). Artificial intelligence in education: A review. *IEEE Access*, 8, 75264–75278.

Chen, X., Xie, H., & Hwang, G.-J. (2020c). A multi-perspective study on artificial intelligence in education: Grants, conferences, journals, software tools, institutions, and researchers. *Computers and Education: Artificial Intelligence*, 1, 100005.

Chen, X., Xie, H., Zou, D., & Hwang, G. J. (2020b). Application and theory gaps during the rise of Artifcial Intelligence in Education. *Computers and Education: Artifcial Intelligence*, 1, 100002.

Chen, X., Zou, D., Xie, H., Cheng, G., & Liu, C. (2022). Two decades of artificial intelligence in education. *Educational Technology & Society*, 25(1), 28–47.

Choung, H., David, P., & Ross, A. (2023). Trust in AI and its role in the acceptance of AI technologies. *International Journal of Human–Computer Interaction*, 39(9), 1727–1739.

Collina, L., Sayyadi, M., & Provitera, M. (2023). Critical issues about AI accountability answered. *California Management Review Insights*, 1–15.

Corritore, C. L., Marble, R. P., Wiedenbeck, S., Kracher, B., & Chandran, A. (2005). Measuring online trust of websites: Credibility, perceived ease of use, and risk. *Proceedings of the Eleventh Americas Conference on Information Systems*, Omaha, NE, USA. 2418–2427.

Cotton, D. R., Cotton, P. A., & Shipway, J. R. (2023). Chatting and cheating: Ensuring academic integrity in the era of ChatGPT. *Innovations in Education & Teaching International*, 61, 228–239.

Cranefield, J., Winikoff, M., Chiu, Y. T., Li, Y., Doyle, C., & Richter, A. (2023). Partnering with AI: The case of digital productivity assistants. *Journal of the Royal Society of New Zealand*, 53(1), 95–118.

Cyranoski, D. (2019). Artificial intelligence is selecting grant reviewers in China. *Nature*, 569(7756), 316–317.

D'Mello, S., Olney, A., Williams, C., & Hays, P. (2014). Gaze tutor: A gaze-reactive intelligent tutoring system. *International Journal of Human-Computer Studies*, 70(5), 377–398.

Dede, C. J. (1988). Probable evolution of artificial-intelligence-based educational devices. *Technological Forecasting and Social Change*, 34(2), 115–133. https://doi.org/10.1016/0040-1625(88)90061-3

Dehouche, N. (2021). Plagiarism in the age of massive generative pre-trained transformers (GPT-3). *Ethics in Science and Environmental Politics*, 2, 17–23.

Delipetrev, B., Tsinaraki, C., & Kostic, U. (2020). *Historical evolution of artificial intelligence*. Analysis of the three main paradigm shifts in AI. European Commission, 1–35.

Dupps Jr, W. J. (2023). Artificial intelligence and academic publishing. *Journal of Cataract & Refractive Surgery*, 49(7), 655–656.

Dwivedi, Y. K., Kshetri, N., Hughes, L., Slade, E. L., Jeyaraj, A., Kar, A. K., … & Wright, R. (2023). "So what if ChatGPT wrote it?" Multidisciplinary perspectives on opportunities, challenges and implications of generative conversational AI for research, practice and policy. *International Journal of Information Management*, 71, 102642.

Feine, J., Gnewuch, U., Morana, S., & Maedche, A. (2020). Gender bias in chatbot design. In *Chatbot Research and Design: Third International Workshop, CONVERSATIONS 2019*, Amsterdam, The Netherlands, November 19–20, 2019, *Revised Selected Papers* 3 (pp. 79–93). Springer International Publishing.

Felzmann, H., Fosch Villaronga, E., Lutz, C., & Tamò-Larrieux, A. (2019). Robots and transparency: The multiple dimensions of transparency in the context of robot technologies. *IEEE Robotics and Automation Magazine*, 26(2), 71–78.

Felzmann, H., Fosch-Villaronga, E., Lutz, C., & Tamò-Larrieux, A. (2020). Towards transparency by design for artificial intelligence. *Science and Engineering Ethics*, 26(6), 3333–3361.

Firat, M. (2023). What ChatGPT means for universities: Perceptions of scholars and students. *Journal of Applied Learning and Teaching*, 6(1).

Følstad, A., Araujo, T., Law, E. L. C., Brandtzaeg, P. B., Papadopoulos, S., Reis, L., … & Luger, E. (2021). Future directions for chatbot research: An interdisciplinary research agenda. *Computing*, 103(12), 2915–2942.

Furey, H., & Martin, F. (2019). AI education matters: A modular approach to AI ethics education. *AI Matters*, 4(4), 13–15.

Gayed, J. M., Carlon, M. K. J., Oriola, A. M., & Cross, J. S. (2022). Exploring an AI-based writing assistant's impact on English language learners. *Computers and Education: Artificial Intelligence*, 3, Article 100055.

Gillath, O., Ai, T., Branicky, M., Keshmiri, S., Davison, R., & Spaulding, R. (2020). Attachment and Trust in Artificial Intelligence. *Computers in Human Behavior*, 115, 106607.

Holmes, W., Bialik, M., & Fadel, C. (2023). *Artificial intelligence in education*. Globethics Publications.

Holmes, W., & Tuomi, I. (2022). State of the art and practice in AI in education. *European Journal of Education*, 57(4), 542–570.

Hwang, G. J., Xie, H., Wah, B. W., & Gašević, D. (2020). Vision, challenges, roles and research issues of Artificial Intelligence in Education. *Computers and Education: Artificial Intelligence*, 1, 100001.

Ismail, I., & Jabri, U. (2023). Academic integrity: Preventing students' plagiarism with Turnitin. *Edumaspul: Jurnal Pendidikan*, 7(1), 28–38.

Jeon, J. (2023). Chatbot-assisted dynamic assessment (CA-DA) for L2 vocabulary learning and diagnosis. *Computer Assisted Language Learning*, 36(7), 1338–1364.

Kay, J., & Luckin, R. (2018). Rethinking learning in the digital age: Making the learning sciences count. In *Proceedings of the 13th International Conference of the Learning Sciences (ICLS)*, London, UK, June (pp. 23–27).

Kirmani, A. R. (2022). Artificial intelligence-enabled science poetry. *ACS Energy Letters*, 8(1), 574–576.

Klein, D. (2011). Why learners choose plagiarism: A review of literature. *Interdisciplinary Journal of E-Learning and Learning Objects*, 7, 97–110.

Kooli, C. (2023). Chatbots in education and research: A critical examination of ethical implications and solutions. *Sustainability*, 15(7), 5614.

Lestari, E. R., & Wicaksono, B. (2023). Exploring students' use of Chatbot AI for solving grammar tasks. *Journal on Education*, 6(1), 9992–10004.

Li, X., Hess, T. J., & Valacich, J. S. (2008). Why do we trust new technology? A study of initial trust formation with organizational information systems. *The Journal of Strategic Information Systems*, 17(1), 39–71.

Li, X., Valacich, J. S., & Hess, T. J. (2004, January). Predicting user trust in information systems: A comparison of competing trust models. In *37th Annual Hawaii International Conference on System Sciences*, 2004. Proceedings of the (p. 10). IEEE.

Li, Y. (2013). Text-based plagiarism in scientific writing: What Chinese supervisors think about copying and how to reduce it in students' writing. *Science and Engineering Ethics*, 19, 569–583.

Liang, W., Tadesse, G. A., Ho, D., Fei-Fei, L., Zaharia, M., Zhang, C., & Zou, J. (2022). Advances, challenges and opportunities in creating data for trustworthy AI. *Nature Machine Intelligence*, 4(8), 669–677.

Liebrenz, M., Schleifer, R., Buadze, A., Bhugra, D., & Smith, A. (2023). Generating scholarly content with ChatGPT: ethical challenges for medical publishing. *The Lancet Digital Health*, 5(3), e105–e106.

Livberber, T., & Ayvaz, S. (2023). The impact of artificial intelligence in academia: Views of Turkish academics on ChatGPT. *Heliyon*, 9(9).

Luan, H., Geczy, P., Lai, H., Gobert, J., Yang, S. J., Ogata, H., … & Tsai, C. C. (2020). Challenges and future directions of big data and artificial intelligence in education. *Frontiers in psychology*, 11, 580820.

Luccioni, A., Bullock, J., Pham, K. H., Lam, C. S. N., & Luengo-Oroz, M. (2020). Considerations, good practices, risks and pitfalls in developing AI solutions against COVID-19. arXiv preprint arXiv:2008.09043.

Luengo-Oroz, M., Bullock, J., Pham, K. H., Lam, C. S., & Luccioni, A. (2021). From artificial intelligence bias to inequality in the time of COVID-19. *IEEE Technology & Society Magazine*, 40(1), 71–79.

Lund, B. D., Wang, T., Mannuru, N. R., Nie, B., Shimray, S., & Wang, Z. (2023). ChatGPT and a new academic reality: Artificial Intelligence-written research papers and the ethics of the large language models in scholarly publishing. *Journal of the Association for Information Science and Technology*, 74(5), 570–581.

Makeleni, S., Mutongoza, B. H., & Linake, M. A. (2023). Language education and artificial intelligence: An exploration of challenges confronting academics in global south universities. *Journal of Culture and Values in Education*, 6(2), 158–171.

Malik, A. R., Pratiwi, Y., Andajani, K., Numertayasa, I. W., Suharti, S., & Darwis, A. (2023). Exploring artificial intelligence in academic essay: Higher education student's perspective. *International Journal of Educational Research Open*, 5, 100296.

Marais, E. (2021). A journey through digital storytelling during COVID-19 Students preparedness to use technology for learning in the language classroom. *Research in Social Sciences and Technology*, 6(2), 169–182.

McKnight, D. H., Choudhury, V., & Kacmar, C. (2002). The impact of initial consumer trust on intentions to transact with a web site: A trust building model. *The Journal of Strategic Information Systems*, 11(3–4), 297–323.

Mesquita, L. D. (2018). The "Artificial" consumer: Approaches between artificial intelligence and marketing. In *ANPAD Meetings–Enanpad* (pp. 3–6).

Mohammed, R. A. A., Shaaban, O. M., Mahran, D. G., Attellawy, H. N., Makhlof, A., & Albasri, A. (2015). Plagiarism in medical scientific research. *Journal of Taibah University Medical Sciences*, 10(1), 6–11.

Mphahlele, A., & McKenna, S. (2019). The use of turnitin in the higher education sector: Decoding the myth. *Assessment & Evaluation in HIgher Education*, 44(7), 1079–1089.

Nemorin, S., Vlachidis, A., Ayerakwa, H. M., & Andriotis, P. (2023). AI hyped? A horizon scan of discourse on artificial intelligence in education (AIED) and development. *Learning, Media and Technology*, 48(1), 38–51.

Nghi, T. T., Phuc, T. H., & Thang, N. T. (2019). Applying AI chatbot for teaching a foreign language: An empirical research. *International Journal of Scientific and Technology Research*, 8(12), 897–902.

Nicholson, J. M., Mordaunt, M., Lopez, P., Uppala, A., Rosati, D., Rodrigues, N. P., … & Rife, S. C. (2021). Scite: A smart citation index that displays the context of citations and classifies their intent using deep learning. *Quantitative Science Studies*, 2(3), 882–898.

Ning, S., & Yan, M. (2010, July). Discussion on research and development of artificial intelligence. In *2010 IEEE International Conference on Advanced Management Science (ICAMS 2010)* (Vol. 1, pp. 110–112). IEEE.

Okuyama, K., & Suzuki, K. (2023). Correlators of double scaled SYK at one-loop. *arXiv preprint*. arXiv:2303.07552.

Pinzolits, R. (2023). AI in academia: An overview of selected tools and their areas of application. *MAP Education and Humanities*, 4, 37–50.

PK, F. A. (1984). What is artificial intelligence? “Success is no accident. It is hard work, perseverance, learning, studying, sacrifice and most of all, love of what you are doing or learning to do”, 65.

Pokrivčáková, S. (2019). Preparing teachers for the application of AI-powered technologies in foreign language education. *Journal of Language & Cultural Education*, 7(3), 135–153.

Popenici, S. A. D., & Kerr, S. (2017). Exploring the impact of artificial intelligence on teaching and learning in higher education. *Research and Practice in Technology Enhanced Learning*, 12(1), 1–13.

Pratt, E. (2017). *Artificial Intelligence and Chatbots in Technical Communication – 19 A Primer*. Tekom (p. 3). Retrieved from http://intelligent-information.blog/wp20content/uploads/2017/09/A-Primer-AI-and-Chatbots-in-T

Qodirov, X. (2023). Ai in translation: Benefits, challenges, and the road ahead. Лучшие интеллектуальные исследования, 6(2), 122–126.

Raja, R., & Nagasubramani, P. C. (2018). Impact of modern technology in education. *Journal of Applied and Advanced Research*, 3(1), 33–35.

Rodrigues, R. (2020). Legal and human rights issues of AI: Gaps, challenges and vulnerabilities. *Journal of Responsible Technology*, 4, 100005.

Roll, I., & Wylie, R. (2016). Evolution and revolution in artificial intelligence in education. *International Journal of Artificial Intelligence in Education*, 26(2), 582–599.

Rudolph, J., Tan, S., & Tan, S. (2023). War of the chatbots: Bard, Bing chat, ChatGPT, Ernie and beyond. The new AI gold rush and its impact on higher education. *Journal of Applied Learning and Teaching*, 6(1).

Schiff, D., Rakova, B., Ayesh, A., Fanti, A., & Lennon, M. (2020). Principles to practices for responsible AI: Closing the gap. arXiv preprint arXiv:2006.04707.

Scollon, R. (1995). Plagiarism and ideology: Identity in intercultural discourse. *Language in Society* 24(1), 1–28.

Shin, D. (2020). User perceptions of algorithmic decisions in the personalized AI system: Perceptual evaluation of fairness, accountability, transparency, and explainability. *Journal of Broadcasting & Electronic Media*, 64(4), 541–565.

Shin, D. (2021). The effects of explainability and causability on perception, trust, and acceptance: Implications for explainable AI. *International Journal of Human-Computer Studies*, 146, 102551.

Siau, K., & Wang, W. (2020). Artificial intelligence (AI) ethics: Ethics of AI and ethical AI. *Journal of Database Management (JDM)*, 31(2), 74–87.

Stahl, B. C., Antoniou, J., Bhalla, N., Brooks, L., Jansen, P., Lindqvist, B., … & Wright, D. (2023). A systematic review of artificial intelligence impact assessments. *Artificial Intelligence Review*, 56(11), 12799–12831.

Sterner, E. (2021). Scite: Providing Additional Context for Citation Statements. *Issues in Science and Technology Librarianship*, 98, 1–4.

Stokel-Walker, C. (2022). AI bot ChatGPT writes smart essays—Should academics worry. *Nature*. https://doi.org/10.1038/d41586-022-04397-7

Stokel-Walker, C., & Van Noorden, R. (2023). What ChatGPT and generative AI mean for science. *Nature*, 614(7947), 214–216.

Studente, S., Ellis, S., & Garivaldis, S. F. (2020). Exploring the potential of chatbots in higher education: A preliminary study. *International Journal of Educational and Pedagogical Sciences*, 14(9), 768–771.

Sullivan, M., Kelly, A., & McLaughlan, P. (2023). ChatGPT in higher education: Considerations for academic integrity and student learning. *Journal of Applied Learning and Teaching*, 6(1), 1–11.

Sutherland-Smith, W. (2005). Pandora's box: Academic perceptions of student plagiarism in writing. *Journal of English for Academic Purposes*, 4(1), 83–95.

Taecharungroj, V. (2023). "What can ChatGPT do?" Analyzing early reactions to the innovative AI Chatbot on Twitter. *Big Data and Cognitive Computing*, 7(1), 35.

Taylor, J., & Kochem, T. (2022). Access and empowerment in digital language learning, maintenance, and revival: A critical literature review. *Diaspora, Indigenous, & Minority Education*, 16(4), 234–245.

Thorat, S. A., & Jadhav, V. (2020, April). A review on implementation issues of rule-based chatbot systems. In *Proceedings of the International Conference on Innovative Computing & Communications (ICICC)*.

Tlili, A., Shehata, B., Adarkwah, M. A., Bozkurt, A., Hickey, D. T., Huang, R., & Agyemang, B. (2023). What if the devil is my guardian angel: ChatGPT as a case study of using chatbots in education. *Smart Learning Environments*, 10(1), 15.

Turing, A. M. (1950). Computing machinery and intelligence. In *The essential turing: The ideas that gave Birth to the computer age*. Ed. B. Jack Copeland. Oxford: Oxford UP, 433–464.

Vargas-Murillo, A. R., de la Asuncion, I. N. M., & de Jesús Guevara-Soto, F. (2023). Challenges and opportunities of AI-assisted learning: A systematic literature review on the impact of ChatGPT usage in higher education. *International Journal of Learning, Teaching and Educational Research*, 22(7), 122–135.

Weizenbaum, J (1976). *Computer power and human reason: From judgement to calculation*. Penguin Books Ltd, Harmondsworth, England.

West, D. M., & Karsten, J. (2015). How robots, artificial intelligence, and machine learning will affect employment and public policy. Retrieved December 28, 2016.

Winke, P. M., & Isbell, D. R. (2017). Computer-assisted language assessment. In S. L. Thorne, & S. May, *Language, Education and Technology* (pp. 313–326). Springer.

Yao, F., Zhang, C., & Chen, W. (2015). Smart talking robot Xiaotu: Participatory library service based on artificial intelligence. *Library Hi Tech*, 33(2), 245–260.

Zhai, X., Chu, X., Chai, C. S., Jong, M. S. Y., Istenic, A., Spector, M., … & Li, Y. (2021). A review of artificial intelligence (AI) in education from 2010 to 2020. *Complexity*, 2021, 1–18.

Zhang, K., & Aslan, A. B. (2021). AI technologies for education: Recent research & future directions. *Computers and Education: Artificial Intelligence*, 2, 100025.

Zhao, J., Wang, T., Yatskar, M., Cotterell, R., Ordonez, V., & Chang, K. W. (2019). Gender bias in contextualized word embeddings. *arXiv preprint arXiv*:1904.03310.

Zhu, J. (2022). AI ethics with Chinese characteristics? Concerns and preferred solutions in Chinese academia. *AI & Society*, 1–14. https://doi.org/10.1007/s00146-022-01578-w

7 Transforming Education through AI-Enhanced Content Creation and Personalized Learning Experiences

M. Shanmuga Sundari and Harshini Reddy Penthala
BVRIT HYDERABAD College of Engineering for Women, India

Anand Nayyar
School of Computer Science, Duy Tan University, Da Nang, Viet Nam

7.1 INTRODUCTION

Education has long been the cornerstone of human civilization, a timeless endeavor that transcends generations. It is a dynamic process that empowers individuals, shapes societies, and drives progress. Yet, the educational landscape is not static; it evolves in response to the shifting sands of time, culture, and technology. The digital age, with its rapid technological advancements, presents education with both unprecedented challenges and transformative opportunities. At the heart of this evolution lies the fusion of education and Artificial Intelligence (AI) [1], a convergence that promises to redefine how we learn, teach, and navigate the ever-expanding realms of knowledge as shown in Figure 7.1.

The integration of Artificial Intelligence into education represents a watershed moment in the evolution of learning and teaching. While the advent of AI in education is a relatively recent development, it has already garnered significant attention and investment from educators, policymakers, and technologists worldwide. The rationale for this integration is multifaceted, grounded in a vision of education that transcends the limitations of traditional approaches. This section delves deeper into the compelling reasons behind the widespread adoption of AI in educational settings.

At the heart of education lies a fundamental aspiration: to foster improved learning outcomes. The traditional educational model, characterized by standardized curricula and instructional approaches, has undoubtedly contributed to the advancement

 DOI: 10.1201/9781032644509-7

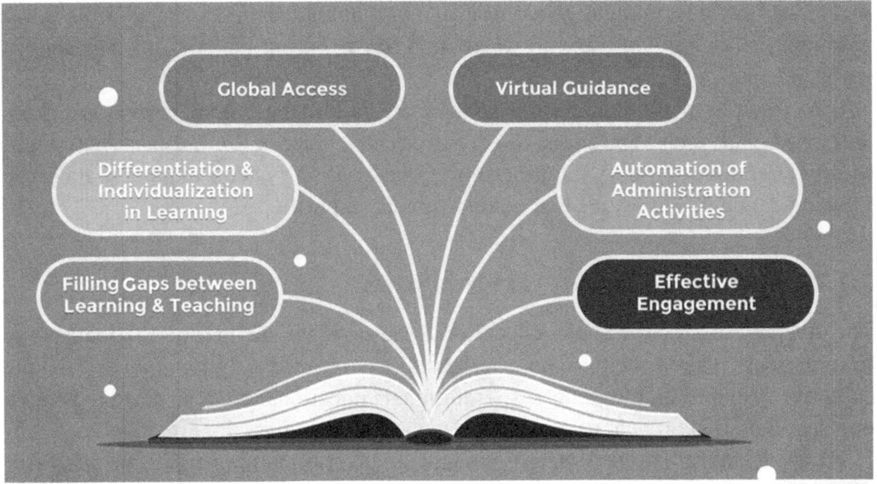

FIGURE 7.1 AI in education.

Source: aiworldschool.com.

of knowledge. However, it has its limitations when it comes to catering to the diverse needs and learning styles of individual students.

AI addresses this challenge head-on by offering a personalized learning experience. Machine learning algorithms can analyze a student's strengths, weaknesses, and learning preferences, creating a tailored educational journey. For example, if a student excels in mathematics but struggles with grammar, AI-powered platforms can adapt content, pacing, and assessments to focus on the areas requiring improvement. This personalized approach increases engagement, motivation, and ultimately, comprehension.

The result is a learning environment that is finely tuned to each student's unique needs and abilities, potentially leading to higher retention rates, increased comprehension, and a deeper love for learning. The shift from a one-size-fits-all to an adaptive, personalized model is a compelling rationale for AI's integration in education.

7.1.1 Accessibility and Scalability

AI is poised to bridge these divides by enhancing the accessibility and scalability of education. With online and AI-driven platforms, learners in remote or underserved areas gain access to high-quality educational resources. These platforms can deliver content, assessments, and even interactive experiences to students worldwide, regardless of their geographical location. The democratization of education through AI ensures that learning is not limited by geographic boundaries.

Moreover, AI can address the challenge of scalability within educational systems. As student populations continue to grow, the burden on educators to provide personalized attention becomes increasingly onerous. AI-powered tools can assist teachers in managing larger cohorts while still offering individualized support.

Automated grading, content generation, and data analytics reduce the administrative workload, allowing educators to focus more on fostering meaningful interactions with students.

The result is a more inclusive educational landscape where individuals from diverse backgrounds, including those historically marginalized, can access high-quality learning experiences. Furthermore, as education systems expand to meet the growing demand, AI ensures that quality remains a priority.

7.1.2 Adapting to Diverse Learning Styles

The diversity of learners in any educational setting [2] is staggering. Students come from various cultural backgrounds, possess distinct learning styles, and have unique cognitive strengths and weaknesses. Traditional educational models, designed with a one-size-fits-all approach, often struggle to cater to this diversity effectively.

AI is a chameleon in this regard. It adapts to diverse learning styles, meeting students where they are and accommodating their individual needs. Whether a student is a visual, auditory, or kinesthetic learner, AI-powered platforms can offer content in formats that resonate with their preferred style. For example, visual learners may benefit from interactive multimedia presentations, while auditory learners might benefit from audio-enhanced content.

Moreover, AI can identify specific areas where a student requires additional support. If a student is struggling with mathematical concepts but excelling in literary analysis, AI can provide targeted resources and exercises to address the areas of challenge. This adaptive approach prevents students from falling behind and reinforces their strengths.

For students with disabilities or special learning needs, AI can be a game-changer. Text-to-speech and speech-to-text technologies, for instance, can break down barriers for students with visual or hearing impairments. AI can generate accessible content that meets the highest standards of inclusivity. In essence, AI's ability to adapt to diverse learning styles ensures that no student is left behind, catering to the unique strengths and needs of each learner.

7.2 AI IN CONTENT CREATION: REVOLUTIONIZING EDUCATIONAL RESOURCES

The field of Natural Language Processing (NLP) has emerged as a transformative force in education. NLP, a branch of Artificial Intelligence focused on the interaction between computers and human language [3], holds the potential to revolutionize how educational content is created, disseminated, and experienced.

At its core, NLP enables computers to understand, interpret, and generate human language in a way that was once thought to be solely within the purview of humans. This capacity has profound implications for educational content, where language is the primary medium of instruction and communication.

One of the most compelling aspects of NLP in education is its ability to tailor content to individual learners' needs as shown in Figure 7.2. Traditional textbooks and lecture formats present a one-size-fits-all approach that often falls short in

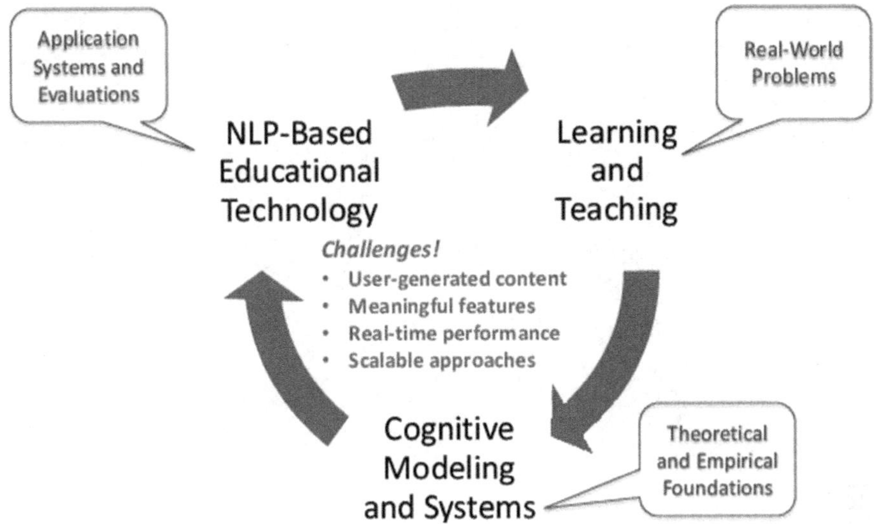

FIGURE 7.2 Natural language processing in education.

Adapted from AAAI Conference 2016, titled "Natural Language Processing for Enhancing Teaching and Learning".

addressing the diverse learning styles, paces, and abilities of students. NLP-driven educational content can adapt in real-time to meet these diverse needs.

7.2.1 Enabling Multilingual and Inclusive Learning

In an increasingly globalized world, multilingualism is a valuable skill. NLP facilitates language learning by providing instant translations, vocabulary suggestions, and pronunciation guides. Learners can explore content in their native language while gradually expanding their proficiency in other languages.

Moreover, NLP plays a pivotal role in making education more inclusive. It can assist students with disabilities, such as dyslexia, by providing text-to-speech capabilities or generating alternative formats, like braille. By removing language and accessibility barriers, NLP ensures that educational content is truly inclusive.

While NLP focuses on the textual aspects of educational content, machine vision takes center stage in reshaping the visual and multimedia dimensions of learning materials. Machine vision, a subset of AI that enables computers to interpret and make sense of visual data, has profound implications for education.

7.2.2 Enabling Augmented Reality and Virtual Reality

Machine vision is integral to the development of Augmented Reality (AR) and Virtual Reality (VR) applications in education. AR overlays digital information onto the real world, while VR immerses learners in entirely digital environments. Both technologies leverage machine vision to create rich, interactive learning experiences.

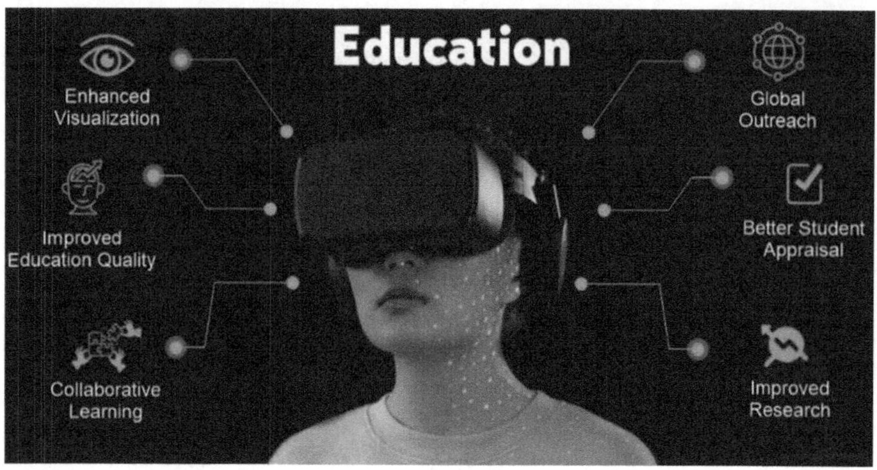

FIGURE 7.3 Benefits of VR in education.

Source: acadecraft.com.

In an AR-enabled biology class, students might use their smartphones or AR glasses to examine a virtual ecosystem superimposed onto a real park. Machine vision identifies plants and animals in the students' view, providing information, quizzes, and interactive simulations. VR, on the other hand, can transport students to historical events, scientific laboratories, or even outer space, offering immersive experiences that deepen understanding and retention as shown in Figure 7.3.

Machine vision can also personalize multimedia content based on individual preferences and needs. For instance, an AI system can analyze a student's facial expressions while viewing educational videos to gauge their engagement. If the student appears disinterested or confused, the AI can adjust the content in real-time, offering explanations or supplementary materials to enhance comprehension.

7.3 PERSONALIZED LEARNING: TAILORING EDUCATION FOR EVERY LEARNER

Adaptive learning platforms stand at the forefront of a pedagogical revolution, offering an educational experience that is as unique as each learner as shown in Figure 7.4. These platforms harness the power of Artificial Intelligence to tailor content and instruction based on individual learners' abilities, preferences, and progress. In this section, we provide a holistic overview of adaptive learning platforms, exploring their architecture, capabilities, and the profound impact they have on education.

Adaptive learning platforms are a dynamic ecosystem that combines AI, educational content, and data analytics. At the core of these platforms are sophisticated algorithms that continuously collect and analyze data. This data includes not only academic performance but also how learners engage with content, the time they spend on specific tasks, and even their emotional responses.

FIGURE 7.4 Adaptive learning examples.

Source: numberdyslexia.com.

These platforms use this data to create a learner profile—a digital representation of each student's strengths, weaknesses, learning preferences, and progress. With this learner profile as a foundation, adaptive learning platforms dynamically adjust the content, pacing, and delivery of instruction to cater to the individual learner.

Adaptive learning platforms go beyond static assessments like quizzes and tests. They continuously assess a student's performance and provide real-time feedback. For example, while a student is working through a physics problem, the platform may identify misconceptions based on the student's responses and offer immediate corrective feedback.

Furthermore, these platforms leverage the power of AI-driven virtual tutors and chatbots to engage in natural language conversations with students. These AI-driven interactions provide a level of support and guidance that is often lacking in traditional classrooms.

7.3.1 A Paradigm Shift in Education

Adaptive learning platforms represent a paradigm shift in education. They recognize that every learner is unique, and the one-size-fits-all approach of traditional education can be limiting. By personalizing learning experiences, these platforms can improve engagement, motivation, and ultimately, learning outcomes.

Data-driven insights form the backbone of personalized learning, offering educators a deep understanding of each learner's progress and needs. In this section, we explore how AI can analyze learner data to create tailored learning pathways that optimize retention and understanding.

The true power of data-driven insights emerges when educators use this information to create personalized learning pathways for their students. These pathways are not static; they adapt and evolve as students' progress and encounter new challenges.

Imagine a high school math teacher using data-driven insights to tailor the curriculum for each student. For a student struggling with geometry, the teacher can assign additional exercises and provide targeted resources. Simultaneously, a student who excels in algebraic concepts can be introduced to more advanced topics. The teacher leverages data to optimize the pacing and content of instruction, ensuring that each student receives precisely what they need to succeed.

7.3.2 Optimizing Retention and Understanding

Data-driven insights also play a crucial role in optimizing retention and understanding. By analyzing how students engage with content, educators can identify patterns and trends. For example, if a particular multimedia resource consistently leads to improved comprehension, educators can incorporate more such resources into their teaching materials.

Furthermore, predictive analytics can help educators anticipate potential roadblocks in a student's learning journey. If a learner exhibits signs of struggling with a particular concept, educators can intervene proactively. They can provide additional support, offer alternative explanations, or connect the student with peer tutors—all based on data-driven insights.

Learner analytics and predictive modeling are indispensable components of personalized learning. In this section, we delve into how AI-driven analytics can anticipate learners' needs and help educators provide timely interventions, ensuring that each student's educational journey is as smooth as possible.

Learner analytics involve the systematic collection, analysis, and interpretation of learner data to inform educational decisions. These analytics offer educators unprecedented insights into each student's progress, strengths, and areas for improvement.

AI-driven learner analytics can provide a comprehensive view of a student's learning journey. They track not only academic performance but also attendance, engagement, and even emotional well-being. For example, a sudden decline in a student's participation or a consistently negative emotional response to certain content can trigger alerts for educators to investigate further.

7.3.3 Predictive Modeling for Early Intervention

Predictive modeling takes learner analytics a step further by anticipating future learning needs based on historical data. These models use machine learning algorithms to identify patterns and trends that may indicate potential challenges or opportunities for each student as seen in Figure 7.5. For instance, if a predictive model detects that a student's performance in a particular subject has been steadily declining, it can alert educators to intervene before the student falls further behind. Early interventions may involve targeted support, additional resources, or modified teaching strategies tailored to the student's specific needs.

Predictive modeling not only helps educators identify areas of concern but also guides them in personalizing support and interventions. These models consider each student's unique learner profile and progress trajectory.

Imagine a scenario where a high school teacher receives a predictive alert indicating that a student is at risk of falling behind in a science course. The teacher can use this information to tailor their approach, providing additional one-on-one support,

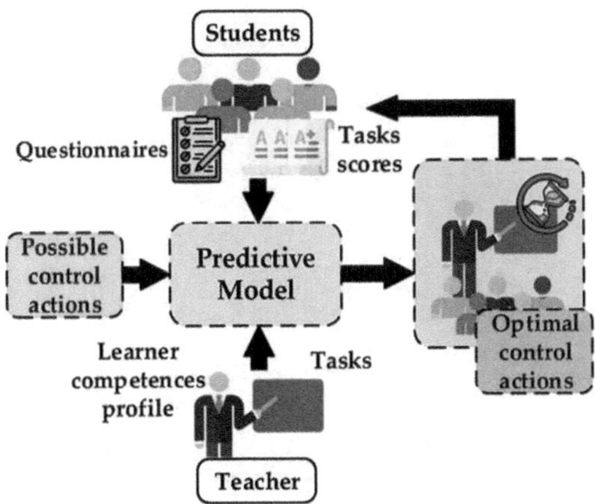

FIGURE 7.5 Predictive model control approach for a specific task.

Adapted from Education Sciences 2023, "A Long-Life Predictive Guidance with Homogeneous Competence Promotion for University Teaching Design".

recommending supplementary materials, or offering study strategies that align with the student's learning preferences.

7.4 THE SYMBIOSIS OF AI AND PEDAGOGY

One of the most significant contributions of AI to pedagogy is personalization. AI-powered systems [4] can analyze individual student profiles and adapt content, pacing, and assessments accordingly. This personalization ensures that each student receives instruction tailored to their unique needs, maximizing engagement and comprehension.

For example, an AI-driven platform can identify that a student excels in mathematics but struggles with writing. It can adjust the curriculum to provide more writing exercises and support, helping the student overcomes their challenges while fostering their strengths.

The collaboration between humans and AI in the classroom represents a harmonious convergence of human expertise and AI capabilities as shown in Figure 7.6. In this section, we explore how educators can leverage AI as a valuable teaching assistant, enhancing the overall educational process.

AI serves as an invaluable assistant to educators, helping them manage administrative tasks, provide personalized support to students, and optimize instructional strategies. With AI-driven grading and assessment tools, educators can save time on routine tasks, such as grading assignments and exams. This time-saving allows educators to allocate more of their energy to meaningful interactions with students.

Moreover, AI-driven virtual tutors and chatbots can engage in natural language conversations with students, providing immediate clarification and feedback. This one-on-one support fosters a more personalized and responsive learning environment.

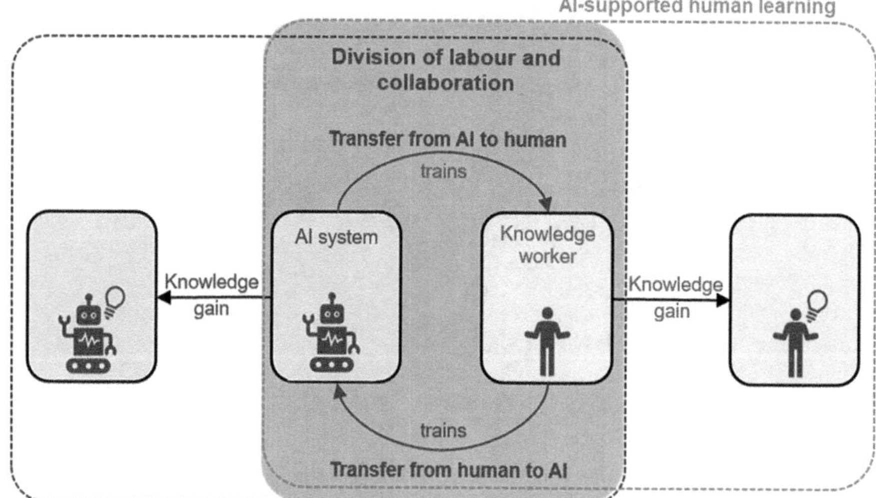

FIGURE 7.6 Collaboration between humans and AI systems

Source: hymeki.

AI can also enhance instructional design by recommending resources, materials, and strategies based on data-driven insights. For instance, an AI system can suggest supplementary readings, multimedia resources, or teaching approaches that align with a particular learning objective or student need.

Furthermore, AI can help educators adapt instruction in real time. If an AI platform detects that a significant portion of the class is struggling with a particular concept, it can signal the educator to provide additional explanations or alternative approaches to ensure comprehension.

7.4.1 AI-Enabled Professional Development

Educators themselves can benefit from AI-enabled professional development. AI systems can analyze an educator's teaching style, strengths, and areas for improvement. Based on this analysis, AI can recommend professional development opportunities, workshops, or resources that align with the educator's goals.

Furthermore, AI can facilitate collaboration among educators by connecting them with peers who have similar interests or expertise. This collaborative learning environment enhances professional growth and knowledge-sharing.

Augmented Reality bridges the physical and digital realms, overlaying digital information onto the real world. In education, AR can bring textbooks, worksheets, and even classrooms to life. For example, students can use AR-enabled devices or apps to explore 3D models of historical landmarks, dissect virtual organisms, or interact with virtual science experiments.

AR also enhances field trips by providing students with contextual information and interactive elements at historical sites, museums, or natural habitats [5]. This technology enables learners to engage with content in a multisensory and interactive manner, deepening their understanding and retention.

7.4.2 VR: Immersive Simulations and Virtual Laboratories

Virtual Reality immerses students in entirely digital environments, providing experiential learning opportunities that were once inaccessible. VR applications in education include immersive history tours, virtual laboratories, and realistic medical simulations as seen in Figure 7.7.

For instance, medical students can practice surgical procedures in a safe and controlled VR environment, enhancing their skills and confidence before working with real patients. Similarly, physics students can explore complex concepts through interactive VR simulations, promoting hands-on learning and experimentation.

AR and VR hold the potential to transform passive learners into active participants. By immersing students in a subject, these technologies boost engagement and foster a deeper understanding of complex topics. Students can explore historical events, visit distant planets, or conduct virtual science experiments—all within the immersive world of AR and VR.

Moreover, AR and VR offer inclusivity benefits. Learners with diverse abilities can engage with content in ways that suit their strengths. For example, students with visual impairments can use AR to access auditory descriptions and tactile feedback, making educational content more accessible.

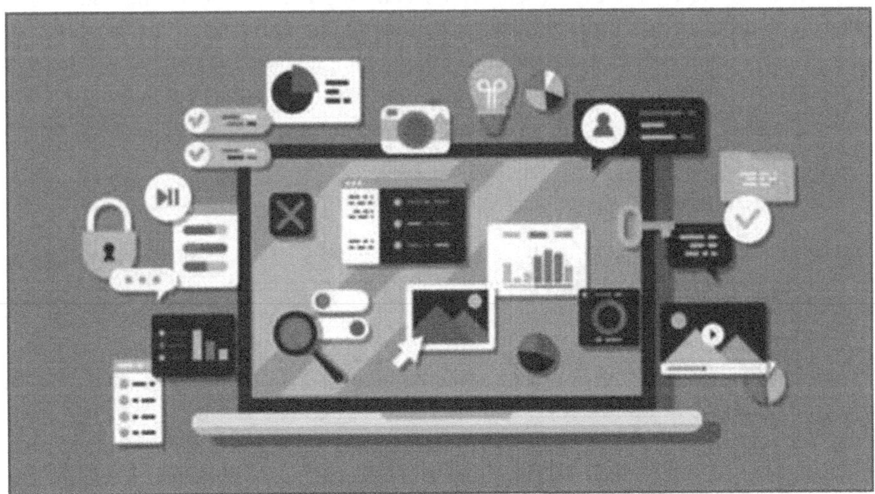

FIGURE 7.7 Virtual reality technology.

Source: pharmacy.howard.edu.

7.5 ETHICAL CONSIDERATIONS AND RESPONSIBLE IMPLEMENTATION

Safeguarding sensitive student data requires a multifaceted approach. Educational institutions must implement robust security measures to protect data from unauthorized access, breaches, or cyberattacks. This includes encryption, secure authentication, and regular security audits.

Furthermore, institutions must establish clear data governance policies, outlining how data is collected, stored, and used. These policies should adhere to relevant data protection regulations, such as the Family Educational Rights and Privacy Act (FERPA) in the United States. Institutions must also educate students, educators, and administrators about their rights and responsibilities regarding data privacy.

One of the paramount ethical considerations in AI-driven education [6] is compliance with data protection regulations. These regulations vary by region but share common principles: the need for informed consent, data minimization (collecting only necessary data), transparency, and the right to access and delete personal data.

Educational institutions must ensure that their AI-enhanced education platforms and practices comply with these regulations. They should also be prepared to respond to data access requests from students and guardians promptly.

7.5.1 Ensuring Inclusivity and Accessibility in AI-Driven Learning

One of the primary ways AI enhances inclusivity is by making educational content and tools accessible to all. For students with disabilities, AI-powered accessibility features can provide alternatives to traditional learning resources. Text-to-speech technology, for example, can read aloud text-based content for students with visual impairments or reading difficulties.

Similarly, AI can generate closed captions and subtitles for videos, making audio-visual content accessible to students with hearing impairments. Furthermore, AI-driven content translation tools can break down language barriers, allowing students with diverse language backgrounds to engage with educational materials in their preferred language.

7.5.2 Universal Design for Learning (UDL)

Universal Design for Learning (UDL) is an educational framework that emphasizes designing educational materials [7] and environments to be accessible to the widest range of learners as seen in Figure 7.8. AI aligns seamlessly with UDL principles by providing multiple means of representation, engagement, and expression.

AI-driven learning platforms can offer content in various formats, such as text, audio, or interactive multimedia, allowing students to choose the format that suits their learning style or accessibility requirements best. Furthermore, AI can provide flexible assessment options, accommodating diverse ways in which students can demonstrate their understanding.

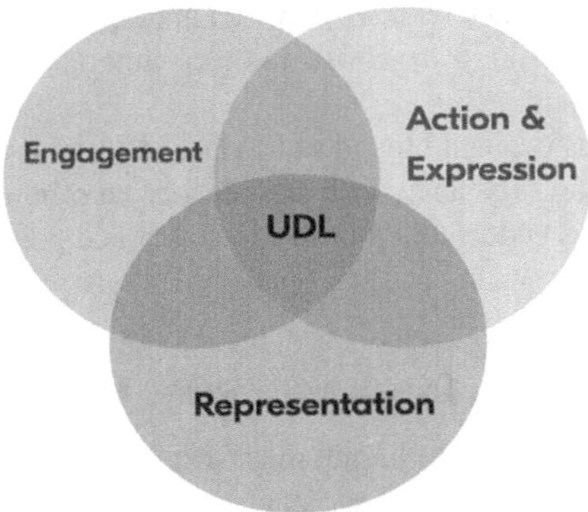

FIGURE 7.8 Universal design for learning.

Source: letsgolearn.com.

7.5.3 Case Studies: Implementing AI in Education

7.5.3.1 Real-World Transformations: AI in Diverse Educational Settings

This takes us on a journey through the diverse landscape of education, showcasing real-world examples of how AI is being successfully implemented across different educational settings. From K-12 schools to higher education institutions, these exemplary use cases illustrate the transformative power of AI in education.

7.5.4 K-12 Schools: Personalized Learning at Scale

In K-12 education as shown in Figure 7.9, AI-driven personalized learning platforms have gained significant traction. These platforms adapt content and pace to individual

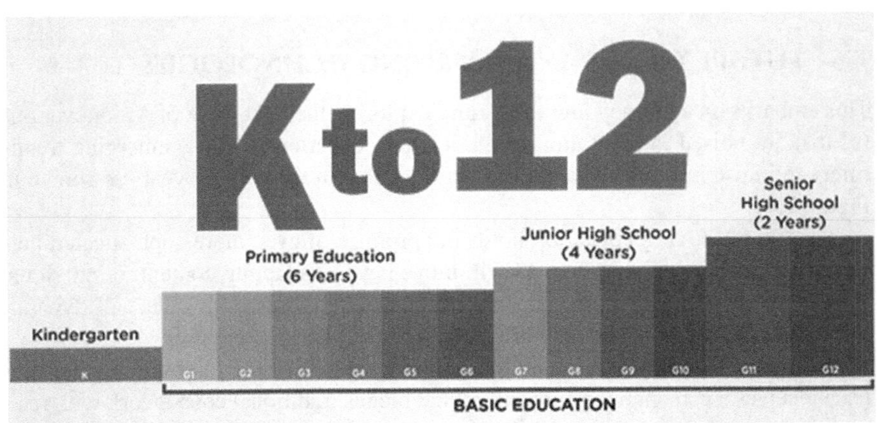

FIGURE 7.9 Learning scale of K-12 school.

Source: philippineeducationhub.com.

student needs, making learning more engaging and effective. A case in point is a school district that implemented an AI-powered math program. It not only improved math scores but also fostered a growth mindset among students who previously struggled with the subject.

Furthermore, AI has been employed in K-12 to identify at-risk students who may need additional support. By analyzing attendance, engagement, and performance data, AI models can alert educators to intervene early and prevent students from falling behind.

7.5.5 Higher Education: Enriching Learning Experiences

Higher education institutions are harnessing AI to enrich learning experiences and support faculty. Chatbots and virtual assistants, powered by AI, are increasingly used to answer student inquiries, provide course information, and even offer academic advising.

In the realm of research, AI-driven analytics [8] are assisting researchers in processing vast datasets, accelerating discoveries, and improving data-driven decision-making. For example, universities use AI to analyze research papers and extract valuable insights, enabling researchers to stay at the cutting edge of their fields.

7.5.6 Corporate Training: Upskilling the Workforce

AI is not limited to traditional education settings; it is also transforming corporate training. Companies are adopting AI-driven platforms to deliver personalized and just-in-time training to employees. These platforms analyze individual skill gaps and deliver targeted content to bridge those gaps.

Moreover, AI-driven simulations and virtual reality are used to train employees in high-risk industries, such as aviation and healthcare. Virtual scenarios provide a safe environment for employees to practice skills and decision-making without real-world consequences.

7.6 FUTURE TRENDS AND EMERGING TECHNOLOGIES

This embarks on a journey into the future, exploring the next wave of AI innovations [9] that are poised to revolutionize the learning experience. These emerging trends offer a glimpse into how education is set to evolve in the coming years as shown in Figure 7.10.

The future of AI in education holds the promise of even more sophisticated personalized learning experiences. AI will move beyond adapting content to individual student needs to create entire personalized curricula. These curricula will dynamically adjust based on a student's progress, interests, and career goals.

Imagine a scenario where a high school student interested in environmental science receives a personalized curriculum that blends traditional coursework with real-world environmental projects. AI-driven analytics continuously evaluate the student's understanding and adapt the curriculum to provide deeper challenges or additional support as needed.

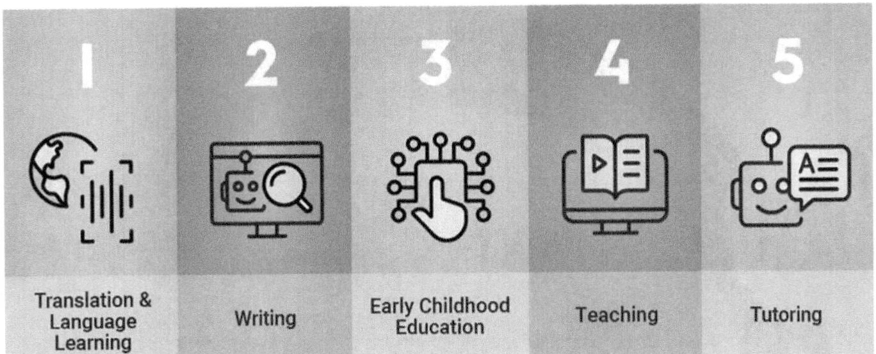

FIGURE 7.10 AI uses in education.

Source: The Motley Fool.

AI will play a pivotal role in fostering collaborative learning. Intelligent virtual classrooms [10] will leverage natural language processing and AI-driven recommendations to facilitate group discussions, peer review, and collaborative projects. AI will identify students with complementary skills and learning styles, forming optimal study groups.

Moreover, AI-driven chatbots and virtual teaching assistants will be equipped with enhanced natural language capabilities, making learning more interactive and conversational. These AI entities will offer instant clarification, guidance, and even facilitate Socratic-style dialogues to deepen students' critical thinking.

The future of assessments will be transformed by AI. Traditional standardized tests will make way for AI-driven, adaptive assessments that provide a more comprehensive view of a student's capabilities. These assessments will evaluate not only knowledge but also critical thinking, problem-solving, and creativity.

Furthermore, AI will enable continuous formative assessments, offering real-time insights into student learning. Educators will receive instant feedback on student performance, allowing for immediate interventions and adjustments to teaching strategies.

7.7 CHALLENGES AND OVERCOMING BARRIERS

This sheds light on the practical challenges and barriers that educators and institutions may encounter when embarking on the journey of implementing AI in education can be seen in Figure 7.11. These challenges encompass various aspects, from technology adoption to cultural shifts.

One of the foremost practical hurdles in AI implementation is the associated costs. AI infrastructure, software development, data management, and training programs can be financially demanding. Many educational institutions, especially smaller ones, may face budget constraints that impede the adoption of AI solutions.

Moreover, securing skilled AI professionals can be a challenge, as there is a growing demand for AI talent across various industries. Competition for qualified AI experts [11] may drive up labor costs.

FIGURE 7.11 Challenges of AI in education.

Source: researchgate.net.

Data privacy and security concerns are paramount in AI implementation. Educational institutions handle sensitive student data, and any breach or mishandling of this data can lead to legal and ethical consequences. Ensuring data privacy and security requires robust systems and policies, which can be complex to implement. Institutions must navigate the intricacies of data encryption, secure storage, access controls, and compliance with data protection regulations.

AI-driven education often relies on technology infrastructure, including high-speed internet access and up-to-date devices. However, students from underserved communities may lack access to these resources, creating a digital divide.

Educational institutions must grapple with how to provide equitable access to AI-enhanced learning experiences. Bridging this gap may involve investments in infrastructure, partnerships with community organizations, or creative solutions like mobile learning platforms.

Resistance to change can be a significant barrier, particularly in educational institutions with established practices and cultures. Educators, administrators, and students may be reluctant to embrace AI-driven changes due to fear of the unknown, concerns about job displacement, or simply a preference for traditional teaching methods.

Overcoming resistance to change requires effective change management strategies, including communication, training, and demonstrating the benefits of AI integration. Fostering a culture of innovation and collaboration can also help shift attitudes.

7.7.1 Navigating the AI Integration Maze: Strategies for Success

While the challenges of AI integration are real, they are not insurmountable. This section provides strategies and recommendations for successfully navigating these complexities and ensuring that AI enhances rather than disrupts education. The below points are added into enhancing AI Education:

- Develop a Clear Vision and Strategy
- Secure Adequate Funding and Resources
- Prioritize Data Privacy and Security
- Build a Skilled Team
- Address Digital Divide
- Engage Stakeholders and Communicate Benefits
- Implement Change Management Strategies
- Monitor and Evaluate Progress
- Foster a Culture of Innovation
- Stay Informed and Adapt

This acknowledges the practical hurdles and challenges educators and institutions may face when implementing AI in education. However, it also emphasizes that with the right strategies and a clear vision, these challenges can be overcome. AI integration has the potential to enhance education, improve student outcomes, and prepare learners for the future. By addressing financial constraints, prioritizing data privacy, bridging access disparities, and fostering a culture of innovation, educational institutions can navigate the complexities of AI integration successfully and reap the benefits of this transformative technology.

7.8 EMPOWERING EDUCATORS: TRAINING AND PROFESSIONAL DEVELOPMENT

This underscores the pivotal role of educator empowerment in the successful integration of AI in education. As AI becomes an integral part of the educational landscape, educators need training and capacity-building programs to equip them with the necessary skills and knowledge as shown in Figure 7.12 to harness AI's potential.

FIGURE 7.12 Professional development skills for modern teachers.

Source: lsaglobal.com.

7.8.1 The Educator's Role in AI-Enhanced Education

Educators are at the forefront of AI-enhanced education. Their role extends beyond traditional teaching to include guiding students through AI-driven learning experiences, interpreting AI-generated insights, and creating dynamic learning environments.

To fulfill these roles effectively, educators require specific competencies:

- **Digital Literacy**: Educators must possess a strong foundation in digital literacy, including proficiency in using AI-driven tools, data analysis, and understanding AI algorithms.
- **Data Interpretation**: They need the ability to interpret data generated by AI to make informed decisions about instructional strategies, identifying areas where students need additional support, and personalizing learning pathways.
- **Curriculum Design**: Educators should be adept at designing curricula that leverage AI to adapt content and assessments, ensuring they align with educational objectives and meet students' diverse needs.
- **Ethical AI Use**: Understanding the ethical implications of AI, including privacy, fairness, and bias, is crucial. Educators must be capable of guiding students in responsible AI use.

Educational institutions and organizations are increasingly recognizing the need to invest in training and capacity-building programs for educators [12]. These programs aim to bridge the AI skills gap and empower educators to harness AI effectively in their teaching practices.

Key components of effective training and capacity-building programs include:

- **Tailored Curriculum**: Designing programs that cater to educators' existing knowledge and skills while progressively introducing AI concepts and applications. Training should be relevant to their subject areas and grade levels.
- **Hands-On Experience**: Providing educators with hands-on experience using AI tools and platforms. Practical exercises and real-world scenarios help educators gain confidence in AI integration.
- **Expert Guidance**: Involving AI experts, data scientists, and experienced AI educators in training programs to provide mentorship and guidance. These experts can clarify doubts, offer best practices, and share their experiences.
- **Collaborative Learning**: Encouraging collaboration and knowledge sharing among educators. Peer learning and communities of practice can facilitate the exchange of ideas and strategies for AI-enhanced teaching.
- **Feedback and Assessment**: Incorporating feedback mechanisms and assessments to gauge educators' progress and understanding of AI concepts. Continuous improvement is vital in training programs.

7.8.2 Empowering Educators through Collaboration and Resources

In this section, we explore the resources and support mechanisms available to educators in the AI-enhanced education landscape. These resources aim to foster collaboration, continuous learning, and a sense of community among educators navigating the evolving educational technology landscape.

Online communities [13] and networks provide valuable platforms for educators to connect, collaborate, and exchange ideas about AI in education. These communities often feature discussion forums, webinars, and resource sharing, creating a vibrant ecosystem for learning and growth.

For example, educators can join AI in Education groups on platforms like LinkedIn, participate in Twitter chats focused on educational technology, or engage in online forums dedicated to AI integration in the classroom. These communities provide a space for educators to seek advice, share success stories, and explore innovative AI applications.

Educator-centric AI tools are designed to assist teachers in creating AI-enhanced learning experiences without requiring extensive technical expertise. These tools often feature user-friendly interfaces, pre-built AI models, and templates that simplify the integration of AI into teaching practices.

For instance, AI-driven chatbots can help educators create interactive and responsive virtual classroom environments. These tools can answer student queries, facilitate discussions, and provide real-time feedback, enhancing the overall learning experience.

Educators can access a wealth of professional development resources dedicated to AI in education. These resources encompass online courses, webinars, workshops, and conferences that cater to educators at various skill levels.

Institutions and organizations often offer AI-specific professional development programs. These programs cover topics such as AI fundamentals, AI ethics, data analysis, and curriculum design with AI integration. Educators can leverage these resources to deepen their AI knowledge and refine their teaching strategies.

Educational institutions play a pivotal role in supporting educators on their AI journey. They can establish AI task forces or committees responsible for researching, implementing, and evaluating AI technologies. These bodies can provide guidance, allocate resources, and facilitate collaboration among educators.

Furthermore, institutions can offer incentives for educators who actively engage in AI integration, such as recognition, awards, or opportunities for leadership roles in AI-driven initiatives.

This underscores the importance of empowering educators through training, capacity-building programs, and access to resources and support networks. Educators are key stakeholders in the successful integration of AI in education, and their proficiency in AI-driven tools and practices is instrumental in creating engaging and effective learning environments.

As AI continues to shape the educational landscape, the collaboration and continuous learning of educators will remain essential. By investing in educator empowerment, educational institutions can leverage the transformative potential of AI to enhance teaching and learning experiences, ultimately benefiting students and preparing them for the challenges and opportunities of the future.

7.9 CONCLUSION—ENVISIONING THE FUTURE OF EDUCATION WITH AI

As we conclude this exploration of the transformative power of AI in education, we find ourselves at the intersection of possibility and responsibility. The journey through the preceding chapter has unveiled a landscape where AI is not just a technological advancement but a powerful force poised to reshape education. In this conclusion, we reflect on the key takeaways and the pivotal role of AI in envisioning the future of education.

7.9.1 Enhanced Teaching and Learning

AI is not a replacement for educators but a valuable assistant. It provides educators with data-driven insights, automates administrative tasks, and facilitates more interactive and engaging teaching methods. Educators are empowered to focus on what matters most—guiding students on their learning journey and fostering critical thinking and creativity.

Additionally, AI-powered educational content is more engaging and adaptive, making learning enjoyable and effective. Multimedia integration, natural language processing, and deep learning models enrich educational materials, making them more interactive and immersive.

7.9.2 Ethical Considerations and Inclusivity

This has highlighted the importance of ethical considerations in AI-driven education. Addressing data privacy, algorithmic bias, and ensuring inclusivity are not optional but essential. Responsible AI use is the foundation of a fair and equitable education system.

By prioritizing ethical AI practices, we can ensure that the benefits of AI are distributed equitably among diverse learner populations. This requires a commitment to transparency, fairness, and continuous monitoring of AI systems for potential bias.

7.9.3 Continuous Learning and Collaboration

Another key takeaway is the necessity of continuous learning for educators and educational institutions. As AI evolves, so must the skills and knowledge of educators. Training programs, professional development, and supportive communities are vital for educators to navigate the complex AI landscape successfully.

Furthermore, collaboration between educators, AI experts, policymakers, and stakeholders is crucial. It fosters innovation, knowledge sharing, and the development of best practices. Together, these collaborations can drive the responsible and effective integration of AI into education.

REFERENCES

1. Polak, Sara, Gianluca Schiavo, and Massimo Zancanaro. "Teachers' perspective on artificial intelligence education: An initial investigation." In *CHI Conference on Human Factors in Computing Systems Extended Abstracts*, pp. 1–7, 2022.
2. Baidoo-Anu, David, and Leticia Owusu Ansah. "Education in the era of generative artificial intelligence (AI): Understanding the potential benefits of ChatGPT in promoting teaching and learning." *Journal of AI* 7, no. 1 (2023): 52–62.
3. Chen, Xieling, Di Zou, Haoran Xie, Gary Cheng, and Caixia Liu. "Two decades of artificial intelligence in education." *Educational Technology & Society* 25, no. 1 (2022): 28–47.
4. Babitha, M. M., Sushma, C., and Gudivada, V. K. (2022). Trends of Artificial Intelligence for online exams in education. *International journal of Early Childhood special Education*, 14(01), 2457–2463.
5. Chandra, J. Vijaya, and Sai Kiran Pasupuleti. "Machine learning methodologies for predicting neurological disease using behavioral activity mining in health care." In *2022 8th International Conference on Advanced Computing and Communication Systems (ICACCS)*, vol. 1, pp. 1035–1039. IEEE, 2022.
6. Tapalova, Olga, and Nadezhda Zhiyenbayeva. "Artificial intelligence in education: AIEd for personalised learning pathways." *Electronic Journal of e-Learning* 20, no. 5 (2022): 639–653.
7. Ouyang, Fan, Luyi Zheng, and Pengcheng Jiao. "Artificial intelligence in online higher education: A systematic review of empirical research from 2011 to 2020." *Education and Information Technologies* 27, no. 6 (2022): 7893–7925.
8. Jadala, Vijaya Chandra, Sai Kiran Pasupuleti, and Pachipala Yellamma. "Deep learning analysis using ResNet for early detection of cerebellar ataxia disease." In *2022 International Conference on Advancements in Smart, Secure and Intelligent Computing (ASSIC)*, pp. 1–6. IEEE, 2022.
9. Shanmuga Sundari, M., Pusarla Samyuktha, Alluri Kranthi, and Suparna Das. "Evaluating performance on COVID-19 tweet sentiment analysis outbreak using support vector machine." In *Smart Intelligent Computing and Applications, Volume 1: Proceedings of Fifth International Conference on Smart Computing and Informatics (SCI 2021)*, pp. 151–159. Singapore: Springer Nature Singapore, 2022.
10. Holmes, Wayne, and Kaśka Porayska-Pomsta, eds. *The ethics of artificial intelligence in education: Practices, challenges, and debates*. Taylor & Francis, 2022.

11. Uunona, Gabriel N., and Leila Goosen. "Leveraging ethical standards in artificial intelligence technologies: a guideline for responsible teaching and learning applications." In *Handbook of research on instructional technologies in health education and allied disciplines*, pp. 310–330. IGI Global, 2023.
12. Whig, Pawan, Arun Velu, and Rahul Ready. "Demystifying federated learning in artificial intelligence with human-computer interaction." In *Demystifying federated learning for blockchain and industrial internet of things*, pp. 94–122. IGI Global, 2022.
13. Su, Jiahong, and Weipeng Yang. "Artificial intelligence in early childhood education: A scoping review." *Computers and Education: Artificial Intelligence* 3 (2022): 100049.

8 Exploring the Interplay of Educational Social Media Usage, Procrastination, and Subjective Well-being in the Context of Industry 5.0 Education

Pallavi Patwari and Aparna Vajpayee
PP Savani University, Gujarat, India

8.1 INTRODUCTION

8.1.1 Industry 5.0 and Social Media Usages

In the dynamic landscape of Industry 5.0, marked by transformative technological advancements, concerns have emerged regarding the psychological impact of online communication and its potential adverse effects on users (Phu and Gow, 2019). A pivotal aspect of this technological evolution is the integration of Industry 5.0 in education, where advanced technologies shape learning experiences (Koç and Turan, 2021). Amidst these innovations, the powerful nature of online communication, particularly through social media, raises concerns about its impact on subjective well-being and its role in facilitating procrastination (Hammad and Awed, 2023).

In the realm of Industry 5.0 in education, various platforms and applications, such as virtual classrooms, collaborative online tools, and educational social media, have become integral parts of the learning environment (Koç and Turan, 2021). These technologies not only facilitate educational interactions but also provide real-time information, news, and collaborative opportunities. However, the constant evolution of these tools, including features like interactive elements and real-time feedback mechanisms, can lead to dependencies, both financial and emotional, among educators and students alike.

The relationship between the usage of educational social media, subjective well-being, and procrastination is a complex interplay in the Industry 5.0 educational

DOI: 10.1201/9781032644509-8

landscape (Gerson et al., 2016). On one hand, these platforms contribute to the expansion of educational networks, fostering self-esteem, and providing avenues for self-presentation (Longobardi et al., 2020). On the other hand, concerns arise regarding problematic usage patterns, which may manifest as addictive behaviors and hinder academic progress (Yang, Asbury, and Griffiths, 2019).

Research in the context of Industry 5.0 education suggests a need to examine the prevalence and correlations of problematic technology use, such as smartphone addiction, among students (Nwosu et al., 2020). Studies have explored how issues like academic anxiety, academic procrastination, and self-regulation are interconnected with problematic technology use. The mediating role of technology addiction in influencing self-regulation and academic outcomes highlights the intricate dynamics at play (Yang, Asbury, and Griffiths, 2019).

The ambivalence in the relationship between the use of educational social media and psychological well-being is a critical aspect to explore in the Industry 5.0 educational setting (Stead and Bibby, 2017). While these technologies can enhance learning experiences, they also pose challenges such as social pressure, potential ostracism, and impaired autonomy (Koç and Turan, 2021). Striking a balance between the benefits and drawbacks of digital communication and educational social media use becomes a regulatory challenge for educators and students alike.

Subjective well-being in the context of Industry 5.0 education encompasses emotional reactions, domain satisfactions, and overall life assessments (Kaur et al., 2021). It is imperative to consider not only life satisfaction but also eudaimonic well-being, emphasizing the pursuit of meaning and purpose in educational endeavors (Gerson et al., 2016). Understanding the subjective nature of well-being becomes crucial in a diverse educational landscape where individuals value different aspects of their learning experiences.

Numerous studies have explored the connection between life satisfaction and problematic technology use in educational settings, yielding varying conclusions (Phu and Gow, 2019). The subjective well-being of individuals may be influenced by the integration of Industry 5.0 technologies in education, with potential impacts on academic outcomes and overall happiness (Kaur et al., 2021). The need for a comprehensive understanding of well-being in the context of Industry 5.0 education prompts the inclusion of diverse measures beyond life satisfaction alone.

In the Industry 5.0 education era, the usage of educational social media among undergraduate students is widespread (Koç and Turan, 2021). However, this increased reliance on digital platforms comes with a dual-edged sword, providing convenience but potentially leading to issues such as addiction, procrastination, and symptoms of depression (Nwosu et al., 2020). Researchers have delved into the factors influencing educational social media addiction and its effects on students' mental health and overall life satisfaction (Longobardi et al., 2020).

Educational social media addiction, akin to problematic social networking sites use, is characterized by salience, withdrawal symptoms, relapse, mood modification, tolerance, and conflict (Nwosu et al., 2020). Excessive usage of educational social media is identified as a significant contributor to illogical procrastination, leading to social and personal challenges in students' daily lives (Yang, Asbury, and Griffiths, 2019). This addiction negatively impacts academic performance, interpersonal relationships, and psychological well-being, underscoring the need for a nuanced understanding of its implications (Stead and Bibby, 2017).

Vajpayee, Devanani, and Sanghani (2023) conducted a research study within the context of Industry 5.0, focusing on 200 young adults (18–25 years old) in India. The average age of the participants was 21.5 years. The study aimed to explore the correlation between professional isolation during the Industry 5.0 era and heightened engagement with smartphones, investigating potential connections between technology dependency, workplace stress, and anxiety. Among the 200 participants, 99% acknowledged possessing a smartphone, and 98% reported having functional internet connections. Vajpayee, Devanani, and Sanghani's (2023) study sheds light on the intricate dynamics between the Industry 5.0 landscape, smartphone utilization, and mental well-being. The results underscore the necessity of devising strategies to foster responsible smartphone usage within the evolving industrial paradigm and to mitigate potential negative impacts on professional welfare.

In a related study, Devanani et al. (2022) delved into the relationship between nomophobia and mental health issues, such as depression, anxiety, and stress. The research employed a cross-sectional and quantitative approach, seeking to unravel the intricate interplay between these variables. The sample comprised 200 young adults (18–25 years old) randomly selected from India, maintaining an equal distribution of 100 men and 100 women. The findings revealed a noteworthy association between individuals perceiving their smartphones as integral to their professional identity and the development of attachment to these devices. This attachment heightened the likelihood of experiencing nomophobia and an increased desire for continuous proximity to their phones. Consequently, this attachment led to heightened levels of stress and anxiety among industry professionals. Devanani et al.'s (2022) study advances our comprehension of smartphone addiction within the context of Industry 5.0 and its association with nomophobia. The findings stress the imperative for further research exploring diverse demographics and the formulation of interventions to address the psychological implications of excessive smartphone use within professional settings.

As Industry 5.0 continues to reshape education, addressing the interplay between educational social media usage, subjective well-being, and procrastination becomes essential (Kaur et al., 2021). Striking a balance that harnesses the benefits of technological advancements while mitigating potential drawbacks is a key consideration for educators, policymakers, and students navigating the evolving landscape of Industry 5.0 in education.

8.1.2 Psychological Parameters

The last two decades have witnessed remarkable technological advances, but the pervasive influence of online communication, particularly through social media platforms, has raised concerns about its psychological impact (Turkle, 2012). Social media has become an integral part of daily life, gaining prominence, especially during pandemics, with platforms like Facebook, Instagram, Snapchat, Twitter, and WhatsApp shaping how individuals connect, create content, and access real-time information. This study seeks to empirically explore the intricate relationship between social networking usage, subjective well-being, and procrastination.

8.1.3 Social Networking Sites Dependency

The rapid growth and constant evolution of social networking sites (SNSs) have introduced various features captivating users across age groups, leading to financial, emotional, and mental dependencies. This chapter aims to investigate how this dependency correlates with subjective well-being and contributes to the phenomenon of procrastination, employing empirical evidence (Devanani et al., 2022).

8.1.4 Social Networking Sites Addiction

Social networking sites addiction has become a focal point for researchers, characterized by salience, withdrawal symptoms, relapse, mood modification, tolerance, and conflict. Excessive usage is identified as a significant contributor to irrational procrastination, resulting in social and personal challenges in daily life. This addiction adversely affects academic performance, interpersonal relationships, and psychological well-being (Vajpayee, Devanani and Sanghani, 2023; Devanani et al., 2022).

8.1.5 Impact of SNS Use on Subjective Well-being

Since 2004, platforms like Facebook have transformed modern communication, becoming integral to daily life and influencing social interactions. This study aims to comprehend the influence of SNS use on subjective well-being, recognizing the pivotal role of social ties in individual happiness. Despite a surge in research, mixed results highlight the need to explore potential mediating or moderating variables influencing this relationship.

8.1.6 Challenges of Smartphone and Social Media Integration

The pervasive integration of smartphones and social media into daily life presents challenges to self-control. The impulse to check social media is considered more addictive than smoking or alcohol consumption, creating a self-control dilemma and negative self-image (Jain, Vajpayee, and Sanghani, 2023). The introduction of the Social Media Self-Control Failure (SMSCF) scale aims to measure everyday self-control failures in social media use, anticipating a moderately strong link between SMSCF and problematic social media use (Vajpayee, Devanani, and Sanghani, 2023), the possibility of being encountered with fake news or misinformation is another threat (Vajpayee, Sheokand, and Sanghani, 2022).

8.1.7 Relationship between Facebook Use and Life Happiness

Exploring the relationship between Facebook use and life happiness is a critical aspect of this study, predicting a negative correlation. Additionally, investigating how personality traits impact the connection between SNS use and subjective well-being emphasizes the need to consider eudaimonic well-being, reflecting the pursuit of meaning and purpose in educational endeavors.

8.1.8 Procrastination and Its Link to Subjective Well-being

Procrastination, characterized by willful task delay despite potential negative outcomes, is a prevalent psychological phenomenon. Recognizing three types of procrastination, this study emphasizes irrational procrastination, deliberate avoidance despite awareness of worsening outcomes, as the most common type. The proposal suggests that greater procrastination will lead to lower subjective well-being.

In instantaneous, this research aims to comprehensively explore the intricate interplay between social networking usage, subjective well-being, and procrastination. By drawing from empirical evidence and incorporating innovative scales like SMSCF, the study seeks to contribute valuable insights into the complex relationships within the evolving landscape of technology and human behavior.

8.2 REVIEW OF LITERATURE

In a study by Muslikah, Mulawarman, and Andriyani (2018), they investigated students' academic procrastination and social media usage intensity within the context of Industry 5.0. The study, conducted at Semarang State University with a population of 419 students, utilized a Random Sampling Technique. The analysis, using the product moment, revealed an inverse correlation between academic procrastination and social media usage intensity. The findings contribute valuable insights into how students leverage social media within the framework of Industry 5.0 to delay academic tasks, offering implications for universities aiming to understand and address this phenomenon (Muslikah, Mulawarman, and Andriyani, 2018).

Similarly, Yang, Asbury, and Griffiths (2019) conducted a study among Chinese university students to explore problematic cell phone usage in the context of Industry 5.0 education. With a sample of 475 students, the research delved into the relationships between problematic smartphone use (PSU), academic anxiety, academic procrastination, self-regulation, and subjective well-being. Path analysis revealed that PSU significantly predicted academic procrastination and anxiety, and self-regulation played a mediating role in these associations. This study sheds light on the intricate connections between technology use, academic outcomes, and psychological well-being in the Industry 5.0 educational landscape (Yang, Asbury, and Griffiths, 2019).

Phu and Gow (2019) investigated Facebook use and its association with subjective happiness and loneliness, considering Industry 5.0's impact on social interactions. With a sample size of 332 participants, regression studies unveiled the influence of Facebook variables on loneliness and subjective happiness. The findings indicated that the number of Facebook friends was a significant predictor of reduced loneliness, emphasizing the role of online social connections in Industry 5.0. Furthermore, persistent use of Facebook was linked to increased loneliness and negative body image (Jain, Vajpayee, and Sanghani, 2023), showcasing the nuanced effects of social media engagement in the evolving digital era (Phu and Gow, 2019).

In a study by Longobardi, Settanni, Fabris, and Marengo (2020), Instagram popularity's association with subjective happiness was explored within the context of

Industry 5.0, considering the mediating roles of cyber victimization and social media addiction. The research focused on middle school students actively using Instagram. The findings highlighted a concerning trend where increased popularity on Instagram correlated with a higher risk of developing behavioral addiction and experiencing cyber aggressions. Industry 5.0's influence on social media usage patterns among young adults underscores potential psychological well-being implications (Longobardi, Settanni, Fabris, and Marengo, 2020).

Nwosu et al. (2020) examined the relationship between social media use, problematic internet behaviors, and academic procrastination among undergraduate students at Nnamdi Azikiwe University in Awka. The study, rooted in Industry 5.0, utilized path analysis to evaluate the proposed model. The results indicated that internet addiction significantly predicted academic procrastination, emphasizing the need to consider the broader spectrum of digital behaviors in understanding academic outcomes in the Industry 5.0 educational landscape (Nwosu, Ikwuka, Onyinyechi, and Unachukwu, 2020).

Shah, Mumtaz, and Chughtai (2017) investigated the frequency of subjective happiness and academic procrastination in medical students within the Industry 5.0 framework. Despite a lack of correlation between subjective happiness and academic procrastination, the study provides insights into the psychological well-being of students in a technologically advanced educational setting (Shah, Mumtaz, and Chughtai, 2017).

In a study by Stead and Bibby (2017), the association between personality, fear of missing out, problematic internet use, and subjective well-being was explored. This study, conducted with 495 participants between 18 and 30 years old, considered Industry 5.0's influence on digital behaviors and their impact on well-being. The findings indicated that problematic internet use and fear of missing out negatively affected interpersonal and emotional connections, highlighting the psychological implications of technology used in Industry 5.0 (Stead and Bibby, 2017).

Gerson et al. (2016) delved into how personality factors influence the relationship between social comparison on Facebook and subjective well-being. This study, set in the Industry 5.0 era, involved 337 respondents and revealed that certain personality traits moderate the impact of Facebook social comparison on subjective well-being. The findings underscore the need to consider individual differences in understanding the psychological effects of social media use in the evolving digital landscape (Gerson et al., 2016).

Kaur et al. (2021) examined the connection between social media tiredness and online subjective well-being, considering Industry 5.0's influence on digital fatigue. With data from 320 participants, the study utilized structural equation modelling to reveal the nuanced relationships between online subjective well-being, social comparison, self-disclosure, privacy concerns, and weariness. The findings contribute to the understanding of how Industry 5.0 shapes individuals' experiences and well-being in the online realm (Kaur et al., 2021).

Koç and Turan (2021) employed structural equation modelling and the Social Cognitive Theory to explore the relationships between personal environment variables (self-esteem and subjective well-being) and behaviors (intensity of SNS use and smartphone addiction). With a sample of 734 business school undergraduate students, the study revealed that young individuals tended to use social networking sites to increase their extrinsic outcome expectations (network size) rather than intrinsic outcomes (subjective well-being). The findings highlight the evolving patterns of digital behaviors in Industry 5.0 and their impact on subjective well-being (Koç and Turan, 2021).

In summary, these studies collectively provide insights into the intricate relationships between technology use, social media engagement, and psychological well-being within the context of Industry 5.0. As technology continues to shape educational and social landscapes, understanding these dynamics becomes crucial for fostering a balanced and positive digital environment.

In the era of Industry 5.0, characterized by the integration of advanced technologies and the transformative impact on various aspects of life, including education and communication, it becomes imperative to explore the need and significance of studying the relationship between social networking usage, procrastination, and subjective well-being.

8.2.1 Need for the Study in Industry 5.0

Influence on Social Dynamics**:** With Industry 5.0 facilitating increased connectivity and collaboration through technological advancements, social networking platforms play a pivotal role in shaping social dynamics. Understanding how these platforms influence users' psychological well-being and procrastination is crucial in adapting to the evolving social landscape.

Technological Dependency**:** As Industry 5.0 promotes a more interconnected and technologically driven society, individuals may become increasingly dependent on social networking platforms. Investigating the impact of this dependency on psychological well-being and procrastination is essential to identify potential challenges and develop strategies for responsible technology use.

Educational Implications**:** In the context of Industry 5.0, where education is significantly influenced by digital technologies, the study can shed light on how students' social networking usage may affect their academic procrastination and overall well-being. This insight is essential for educators and institutions to design effective and mindful digital learning environments.

8.2.2 Significance of the Study in Industry 5.0

Balancing Connectivity and Well-Being**:** Industry 5.0 emphasizes seamless connectivity, but an imbalance in social networking usage may lead to adverse effects on well-being and increased procrastination. Understanding this balance is crucial for individuals, educators, and policymakers to promote responsible and mindful technology use.

Digital Well-Being Strategies**:** The findings of the study can contribute to the development of strategies and interventions to enhance users' digital well-being in the Industry 5.0 landscape. This includes promoting healthy online habits, providing digital literacy education, and creating awareness about the potential impact of excessive social networking usage.

Informing Technological Design**:** For industries involved in the development of social networking platforms and related technologies, the study's insights can inform the design process. It can contribute to the creation of features that prioritize users' mental health, mitigate procrastination tendencies, and foster a positive online environment.

***Optimizing Educational Technology*:** In the educational sector, understanding the relationship between social networking usage, procrastination, and subjective well-being can inform the design and implementation of educational technologies. This knowledge can contribute to the development of platforms that enhance learning experiences without negatively impacting students' well-being.

***Policy Development*:** Industry 5.0 necessitates the development of policies that address the ethical and psychological implications of technology use. The study's findings can contribute valuable information to policymakers, enabling them to create regulations that foster a healthy digital environment while encouraging innovation and connectivity.

In conclusion, investigating the relationship between social networking usage, procrastination, and subjective well-being in the context of Industry 5.0 is vital for adapting to the changing technological landscape. The study's insights can guide individuals, educators, industry professionals, and policymakers in fostering a digitally connected society that prioritizes both technological advancements and the well-being of its users. Studies have also showed that misinformation or fake information being shared by social media also negatively impact the cognitive behavior of the individual (Vajpayee, Sheokand, and Sanghani, 2022).

8.3 METHODOLOGY

Investigating the dynamics of social networking usage, procrastination, and subjective well-being in the context of Industry 5.0.

8.3.1 Objectives of the Study

Building upon insights from existing research, the study aims to:

1. Uncover the relationship between social networking usage and procrastination in the adult population within the Industry 5.0 landscape.
2. Examine the relationship between social networking usage and subjective well-being in the adult population, considering the nuances of Industry 5.0.
3. Investigate the relationship between procrastination and subjective well-being in the adult population within the context of Industry 5.0.

8.3.2 Hypotheses

Based on the stated objectives, the study formulates the following hypotheses:

1. Increased social networking usage will correlate with heightened procrastination among adults.

2. Elevated social networking usage will be associated with decreased subjective well-being in the adult population.
3. Procrastination, when pronounced, will be linked to reduced subjective well-being in the adult demographic.

8.3.3 Operational Definitions of the Variables

Procrastination: Purposely delaying planned actions despite anticipating adverse outcomes due to the delay, as articulated by Steel (2007). Habitual procrastinators tend to avoid challenging tasks and may actively seek distractions.

Social Networking Usage: The extent of engagement with websites and applications that facilitate connection, communication, information sharing, and relationship building. The measurement involves assessing the amount of time an individual spends on these platforms.

Subjective Well-being (SWB): Scientifically referred to as happiness and life satisfaction, SWB encapsulates an individual's perception of their life going well, reflecting how they experience and evaluate their lives and specific domains and activities within it.

8.3.4 Variables

Independent Variable: Social networking usage.

Dependent Variables: Procrastination and subjective well-being.

Control Variable: Age, educational qualification, and mental health condition of participants.

8.3.5 Inclusion Criteria

1. Proficiency in English.
2. Absence of diagnosed mental health conditions.
3. Active engagement in social networking usage.
4. Categorization as adults.

8.3.6 Exclusion Criteria

1. Limited proficiency in English.
2. Diagnosis of mental health conditions.
3. Inactivity in social networking usage.
4. Falling outside the defined age range for adults.

8.3.7 Research Design

The study engaged a sample of 100 individuals encompassing diverse genders within the Indian adult population. The sampling technique employed was random convenience sampling, with 110 responses initially collected and outliers subsequently excluded, leaving 100 responses for data analysis.

8.3.8 Tools of Assessment

1. **Procrastination Scale**: Developed by Lay (1986), this 20-item questionnaire employs a 5-point Likert scale to assess procrastination tendencies. Items 3, 4, 6, 8, 11, 13, 14, 15, 18, and 20 are negatively scored, demonstrating high reliability ($\alpha = 0.876$).
2. **Social Networking Usage Scale**: Developed by Gupta and Bashir (2018), this 19-item self-report questionnaire gauges the frequency of social networking usage using a 5-point Likert scale. The scale exhibits good internal reliability ($\alpha = 0.830$).
3. **Subjective Well-being Inventory**: Developed by Sell (1994), this 40-item questionnaire evaluates subjective well-being using a 3-point Likert scale. The minimum and maximum scores achievable are 40 and 120, respectively.

8.3.9 Procedure of Administration

Participants were approached through phone calls and face-to-face interactions. Following their voluntary consent, a detailed explanation of the study was provided. Participants were then sent a Google form containing the three questionnaires and demographic details. Post-completion, participants received a debriefing, elucidating the significance and application of the study. Questionnaires were scored using the prescribed methods.

8.3.10 Ethical Consideration

Informed Consent: Participants provided informed consent before participation.
Voluntary Participation: Participants had the autonomy to participate or withdraw at any point.
No Harm: Participants faced no physical or psychological harm.
Confidentiality: Data confidentiality was ensured.
Anonymity: Participants' identities were protected.

In conclusion, this methodological framework aims to delve into the intricate dynamics of social networking usage, procrastination, and subjective well-being within the realm of Industry 5.0. By employing established tools of assessment and considering ethical imperatives, the study aspires to contribute nuanced insights into the evolving relationship between technology and human behavior in the contemporary socio-technological landscape.

8.4 RESULTS

8.4.1 Findings in the Context of Industry 5.0: Unravelling the Nexus of Procrastination, Social Networking Usage, and Subjective Well-being

In the pursuit of understanding the intricate relationship between procrastination, social networking usage, and subjective well-being among adults within the context of

Industry 5.0, the study delved into a comprehensive analysis. Findings, gleaned from a sample of 100 participants, were subjected to descriptive statistics and correlational analyses, shedding light on the dynamics within this evolving industrial landscape.

8.4.2 Descriptive Analysis

The descriptive statistics (Table 8.1.) provide a snapshot of the mean and standard deviation (SD) for procrastination, social networking usage, and subjective well-being.

The mean score for procrastination, assessed by the Procrastination Scale (AAS), was 53.70 with an SD of 11.02. Social networking usage, gauged by the Social Networking Usage Scale, yielded a mean of 59.13 and an SD of 15.70. Subjective well-being, measured by the Subjective Well-being inventory, obtained a mean score of 89.20 with an SD of 14.54.

8.4.3 Correlational Analyses

The correlational analyses (Tables 8.2–8.4) unearthed significant insights into the interplay between the variables, providing a nuanced understanding within the paradigm of Industry 5.0.

8.4.4 Procrastination and Social Networking Usage

Correlation is significant at the 0.05 level (2-tailed).

The correlation analysis (Table 8.2) revealed a significant positive relationship ($r = 0.634$, $p > 0.05$) between procrastination and social networking usage. This implies that as procrastination levels increase, social networking usage tends to rise and vice versa. The positive correlation validates Hypothesis 1, suggesting that higher social networking usage leads to more procrastination in the context of Industry 5.0.

TABLE 8.1
Descriptive Analysis of Procrastination, Social Networking Usage, and Subjective Well-being

	N	Mean	SD
Procrastination	100	53.70	11.02
Social Networking Usage	100	59.13	15.70
Subjective well-being	100	89.20	14.54

TABLE 8.2
Correlational Analyses between Procrastination and Social Networking Usage (N = 100)

Variable	1	2
Procrastination	—	0.634
Social networking usage	0.634	—

TABLE 8.3
Correlational Analyses between Social Networking Usage and Subjective Well-being (N = 100)

Variable	1	2
Social networking usage	—	–0.276
Subjective well-being	–0.276	—

TABLE 8.4
Correlational Analyses between Procrastination and Subjective Well-being (N = 100)

Variable	1	2
Procrastination	—	–0.436
Subjective well-being	–0.436	—

8.4.5 Social Networking Usage and Subjective Well-being

Correlation is significant at the 0.05 level (2-tailed).

The analysis (Table 8.3) unveiled a significant negative relationship ($r = -0.276$, $p > 0.05$) between social networking usage and subjective well-being. This indicates that elevated social networking usage is associated with lower levels of subjective well-being. The negative correlation supports Hypothesis 2, suggesting that higher social networking usage leads to lower subjective well-being within the dynamics of Industry 5.0.

8.4.6 Procrastination and Subjective Well-being

Correlation is significant at the 0.05 level (2-tailed).

The correlation analysis (Table 8.4) demonstrated a significant negative relationship ($r = -0.436$, $p > 0.05$) between procrastination and subjective well-being. This signifies that increased procrastination is linked to lower levels of subjective well-being. The negative correlation aligns with Hypothesis 3, indicating that greater procrastination leads to lower subjective well-being among adults in the context of Industry 5.0.

8.5 DISCUSSIONS

8.5.1 Analysis in the Context of Industry 5.0: Decoding the Dynamics of Procrastination, Social Networking Usage, and Subjective Well-being

In the realm of Industry 5.0, this study serves as a compass, navigating the intricate relationships between procrastination, social networking usage, and subjective well-being among adults. The analysis unfolds against the backdrop of an evolving industrial landscape characterized by the integration of smart technologies, interconnected systems, and a digitally savvy workforce.

8.5.2 Descriptive Overview

Table 8.1 encapsulates the descriptive data for procrastination, social networking usage, and subjective well-being, offering a foundational snapshot of the variables in focus. The study's scope resonates with the broader discussions in the literature, aligning with findings that highlight the adverse impacts of excessive social media use on procrastination (Hammad and Awed, 2023) and the mediating role of problematic smartphone use in academic procrastination (Yang, Asbury, and Griffiths, 2019).

8.5.3 Correlational Insights

Tables 8.1–8.4 illuminate the interconnectedness of these variables, drawing correlations that reverberate within the context of Industry 5.0.

i. **Social Networking Usage and Procrastination**: The results affirm a significant positive relationship between social networking usage and procrastination, echoing the findings of previous research (Muslikah, Mulawarman, and Andriyani, 2018). In Industry 5.0, where digital connectivity is ubiquitous, individuals may resort to social platforms as a procrastination outlet, potentially impacting productivity.
ii. **Social Networking Usage and Subjective Well-being**: The study unravels a noteworthy negative correlation between social networking usage and subjective well-being. This finding aligns with the overarching discourse on the detrimental effects of excessive internet use on subjective well-being within the Industry 5.0 milieu (Stead and Bibby, 2017). It raises pertinent questions about achieving a balance between digital connectedness and individual well-being in the contemporary workplace.
iii. **Procrastination and Subjective Well-being**: The analysis underscores a significant negative relationship between procrastination and subjective well-being. This aligns with the broader understanding that procrastination, especially fueled by digital distractions, can contribute to a diminished sense of well-being (Berber Çelik, and Odaci, 2022). The implications for Industry 5.0 underscore the importance of mitigating procrastination tendencies to foster a healthier work environment.

8.5.4 Implications for Industry 5.0

In the era of Industry 5.0, where human-machine collaboration is integral, these findings bear implications for workforce dynamics and organizational strategies:

i. ***Digital Distractions and Productivity***: The positive correlation between social networking usage and procrastination urges organizations to address digital distractions. Industry 5.0, with its emphasis on efficiency and connectivity, necessitates strategies to mitigate procrastination, ensuring optimal productivity in a digitally integrated workplace.

ii. ***Balancing Connectivity and Well-being*****:** The negative correlation between social networking usage and subjective well-being underscores the importance of fostering a balanced approach to digital connectivity. Industry 5.0 can benefit from initiatives that promote mindful technology use, emphasizing employee well-being alongside technological integration.

iii. ***Proactive Well-being Measures*****:** Acknowledging the negative correlation between procrastination and subjective well-being, Industry 5.0 organizations should implement proactive measures to address procrastination tendencies. This may involve cultivating a work culture that encourages time management and provides support systems to enhance employee well-being.

8.6 LIMITATIONS AND FUTURE DIRECTIONS IN INDUSTRY 5.0

While offering valuable insights, this study is not without limitations. The sample size, restricted to young adults in a specific geographic location, may limit the generalizability of results to the diverse workforce of Industry 5.0. Future research in this industrial context should explore clinical samples and consider additional variables such as mental health conditions to enrich the understanding of these intricate relationships. Human cognition has unique relationship with artificial intelligence (Vajpayee and Ramachandran, 2019) that need to be implemented with positive proportions with tailored intervention programs (Vajpayee, Devanani, and Sanghani, 2023).

REFERENCES

Berber Çelik, Ç., and Odaci, H. Subjective well-being in university students: What are the impacts of procrastination and attachment styles? *British Journal of Guidance & Counselling* 50, no. 5 (2022): 768–781.

Devanani, S., Vajpayee, A., and Sanghani, P. Neurological syndrome of anxiety and depression as an outcome of nomophobia. *Journal of Pharmaceutical Negative Results* 13, Special Issue 7 (2022): 7768–7774.

Gerson, J., Plagnol, A., and Corr, P. J. Subjective well-being and social media use: Do personality traits moderate the impact of social comparison on Facebook? *Computers in Human Behavior* 63 (2016): 813–822. doi: 10.1016/j.chb.2016.06.023

Gupta, S., & Bashır, L. (2018). Social Networking Usage Questionnaire: Development and Validation in an Indian Higher Education Context. *Turkish Online Journal of Distance Education*, 19 (4): 214–227. https://doi.org/10.17718/tojde.471918

Hammad, M., and Awed, H. S. The use of social media and its relationship to psychological alienation and academic procrastination. *International Journal of Membrane Science and Technology* 10, no. 2 (2023): 332–340.

Jain, E., Vajpayee, A., and Sanghani, P. Beauty ideals and beyond: understanding the impact of negative body image on adolescent self-esteem and interventions. *Korea Review of International Studies* 16, no. 47 (2023): 44–67.

Kaur, P., Islam, N., Tandon, A., and Dhir, A. Social media users' online subjective well-being and fatigue: A network heterogeneity perspective. *Technological Forecasting and Social Change* 172 (2021): 121039.

Koç, T., and Turan, A. H. The relationships among social media intensity, smartphone addiction, and subjective wellbeing of Turkish college students. *Applied Research in Quality of Life* 16, no. 5 (2021): 1999–2021.

Lay, C. At last, my research article on procrastination. *Journal of Research in Personality* 20, no. 4 (1986): 474–495.

Longobardi, C., Settanni, M., Fabris, M. A., and Marengo, D. Follow or be followed: Exploring the links between Instagram popularity, social media addiction, cyber victimization, and subjective happiness in Italian adolescents. *Children and Youth Services Review* 113 (2020): 104955.

Muslikah, M., Mulawarman, D., and Andriyani, A. Academic procrastination and social media usage intensity among university students. *International Journal of Educational Management and Innovation* 1, no. 2 (2018): 1–10.

Nwosu, L. A., Ikwuka, U. D., Onyinyechi, U. O., and Unachukwu, G. C. Social media use and problematic internet behaviours predict academic procrastination in Nigerian undergraduates. *Frontiers in Psychology* 11 (2020): 2027.

Phu, B., and Gow, A. J. Facebook use and its association with subjective happiness and loneliness. *Computers in Human Behaviour* 92 (2019): 151–159.

Sell, H. (1994). The Subjective Well-Being Inventory (SUBI). *International Journal of Mental Health*, *23*(3), 89–102. http://www.jstor.org/stable/41344695

Shah, S. I. A., Mumtaz, A., & Chughtai, A. S. (2017). Subjective Happiness and Academic Procrastination among Medical Students: The Dilemma of Unhappy and Lazy Pupils. PRAS 1: 008.

Stead, H., and Bibby, P. Personality, fear of missing out and problematic internet use and their relationship to subjective well-being. *Computers in Human Behaviour* 76 (2017): 534–540.

Steel, P. (2007). The nature of procrastination: a meta-analytic and theoretical review of quintessential self-regulatory failure. *Psychological bulletin*, 133 (1): 65–94. doi: 10.1037/0033-2909.133.1.65

Turkle, S. *Alone together: Why we expect more from technology and less from each other*. New York: Basic Books, 2012.

Vajpayee, A., Devanani, S., and Sanghani, P. Social isolation, self-aloofness during Covid-19 and its impact on mobile phone addictions. *Korea Review of International Studies* 16, no. 45 (2023): 60–70.

Vajpayee, A., Sheokand, U., and Sanghani, P. Vulnerability of fake news in the human mind and cognition. *NeuroQuantology* 20, no. 22 (2022): 1406–1413. doi: 10.48047/nq.2022.20.22. NQ10119. eISSN 1303-5150

Vajpayee, A., and Ramachandran, K.K. Reconnoitring artificial intelligence in knowledge management. *International Journal of Innovative Technology and Exploring Engineering (IJITEE)* 8, no. 7C (2019): 114–117.

Yang, Z., Asbury, K., and Griffiths, M. D. An exploration of problematic smartphone use among Chinese university students: Associations with academic anxiety, academic procrastination, self-regulation and subjective wellbeing. *International Journal of Mental Health and Addiction* 17 (2019): 596–614.

9 Advancements and Challenges in Fraudulent Message Detection

Nawnit Kumar
Sri Guru Tegh Bahadur Institute of Management & IT, New Delhi, India

Shalabh K. Mishra
Bharati Vidyapeeth's College of Engineering, New Delhi, India

Ramesh Nuthakki
Atria Institute of Technology, Bengaluru, India

9.1 INTRODUCTION

Financial fraud is a major problem for any institutions and individuals around the world. Fraudulent activities can result in substantial losses for both the banks and their customers. Such activities are often performed by sending fake SMS seeking OTP, and unauthorized links for verifying KYC, etc. Hackers often send SMS or calls to the victims and tempt them by providing, free services, rewards, and other benefits. Since sending messages through SMS is within reach as it does not need any internet connectivity, many people opt for it. Because of this, financial fraud is increasing predominantly. It appears to have several ways of lessening or solving this SMS fraud; however, they are yet to be upgraded. In order to put a stop to these fake text messages, several Android apps are made available on the Play Store, but a majority of people are not aware of them. Sequentially, the existent tracking solutions mainly focus on spam messages, even as email being the major problem, though there is an increase in the usage of these Android apps, SMS spam has turned out to be a great cause of concern. As mobile phones contain personal data including credit/debit card details, usernames, and passwords, SMS being the most inexpensive way for communicating and is also considered to be the simplest way to execute phishing scams. Phishers are scheming the newest ways to obtain this information from smartphones by using SMS as it is direct and effortless. In the recent times, hacking via SMS or phishing type of cyberattacks are increasing predominantly. Phishing using SMS has become very common around the globe where the attacker sends a malicious link to the victim by SMS and tricks the person into clicking on it to steal the details of the victim through the cell phone. To detect communications fraud, many screening

 DOI: 10.1201/9781032644509-9

technologies are in place today which include smart cards, machine learning, user fingerprinting, data matrix scanners, and identification-related systems. Scamsters purchase the contact information of any mobile number through some middlemen and use this to send fraudulent messages creating difficulty in tracking down the criminal. Several works have been done in the area of telephonic fraud (Sahin, Francillon, Gupta, & Ahamad 2017; Zhou, Chen, & Zhou 2015; Xing, Yu, Wang, Zhang, & Ding 2020; Ji, Ma, Li 2017; Li, Zhang, Tuo, & Chang 2018; Sanver & Karahoca 2016).

The main purpose of this proposed method should be to best utilize the machine learning techniques in identifying fraud SMSs. To categorize, we have used a feature set of characteristics that helps in differentiating a fraudulent SMS from the real one. These machine learning techniques have been proven to be effective in evading zero-day exploits and offering high levels of security in filtering spam emails. For smartphones too the same technique is being used to curb spam text messaging. Nevertheless, the features relating to spam detection would be different compared to the spamming of email as text messages are shorter in size, also the customers employ precise language for texting.

9.2 LITERATURE SURVEY

Nowadays, Artificial Intelligence-based techniques are widely used in several areas including fraud detection (Mianning, Li, Li, Zhu & Si 2023; Nuthakki, Aameen, Kumar & Mishra 2023; Nuthakki, Masanta & Yukta 2021, 2022). In this chapter, identifying short messaging fraud by the application of machine learning techniques has been looked into in order to screen spam texts based upon the feature selection. El-Alfy and AlHasan have suggested a short messaging filtering technique for both SMS and email (El-Alfy & Al Hasan 2016). To bring forth extracted features, they both have contemplated various techniques that would lessen the complexity. Along with the two classification methods, Naive Bayes and support vector machine, they utilized a total of 13 characteristics which include probable spam terms, JavaScript code, capital letters, URLs, emotional icons, text data, sender address, wrinkly words, junk mail domain, word structure, and topic field. For testing the proposed method, they have utilized five email and SMS datasets. For screening fraud communications, a message subject structure (MTM) has been suggested by Jialin et al. (2016). In order to define spam correspondence, the message predictive model (MTM) utilizes symbols, context terms, words as well as subject terms. It is postulated upon probability guessing through latent semantic analysis. On classifying SMS spam messages into the unpredictable asymmetry divisions and later on gathering all the phishing of SMS into a distinct report for detecting the keyword patterning utilizing the k-means algorithm. The feature reweighting approach and the great word assault are two strategies for SMS spam detection that have been described by Chan et al. (2015). The two methods used put their focus on both the size and weight of the text and the great word method concentrates on misguiding the classifier's outcome through the use of a minimal count of characters available but on the other hand, the characteristic reweighting method contains an innovative parameterized procedure aimed at resampling the dimensions. For evaluating the analysis, two datasets have been utilized: SMS plus comments. Delany et al.

(2012) discussed the many techniques available for SMS spam filtering as well as the issues related to dataset gathering. To estimate a large dataset containing SMS spam that includes money, contests, dating, rewards, music, services, claim chatting, and voice etc., they have employed ten groups. SMS spam messages were identified through information characteristics by Xu et al. (2012). Two classification techniques have been used by the researchers, SVM and K-nearest neighbors (KNN), and a feature set of factors; static along with temporal for the experiment. It has been observed that with the integration of temporal and network characteristics, spam SMS can be detected exactly and proficiently. In addition to this, they discovered techniques that filter SMS spam messages with the usage of characteristics containing charts and time series thereby excluding the content of the text. Text classification methods were tested on separate cell devices to see how well they filtered SMS spam by Nuruzzaman et al. (2012). The entire learning, screening, and also updating operations have been performed over a different mobile. The proposed method is capable enough in finding out the solution to this problem with exceptional accuracy and lesser memory. Nuruzzaman et al. have made significant utilization and sufficient batch processing without requiring a huge dataset or any machine help (Nuruzzaman, Lee, Abdullah, & Choi 2012).

Uysal et al. presented a technique for SMS spam filtering that employs different feature selection methodologies, namely chi-square metrics and mutual information, to select information (Uysal, Gunal, Ergin, & Gunal 2012). In accordance with the Android platform, they designed a hands-on mobile application for detecting spam texts using two Bayesian-related classification techniques, namely Bayesian and binary. As stated by the authors, the proposed method identifies spam as well as original messages with profound correctness. For SMS spam detection, Yadav et al. created a model text Assassin (Yadav, Kumaraguru, Goyal, Gupta, & Naik 2011). A minimum feature set of 20 features together with a couple of machine learning algorithms known as support vector machine: SVM and Bayesian learning have been used by them. For over two months, a total of 2,000 messages from customers have been collected. As per their approach, if each time, a customer gets a text on mobile, without the user's knowledge, SMS Assassin acquires it, repossesses feature values, and transfers it to the server. In case the messages are stated spam, the consumer cannot open and those will be directed to the spam folder. Jun et al. explained the spam detection approach for secure mobile message communication using machine learning algorithms (Jun, Nazir, Khan, & Ul Haq 2020).

9.3 PROCEDURE

The first and foremost purpose of the proposed method is to categorize fraudulent texts promptly after receiving them on a smartphone solely whether or not they are newly generated spam messages. It started off by collecting the data and concluding the features for our analysis. Next to the finalization of the characteristics, we made a feature representation by picking features from the messages: fair and fraud. The different features that were extracted are utilized in training and testing. In our proposed method, the judgment depends on a few selected characteristics. In the process of training, the extracted features associated with fraudulent messages are utilized to produce a classifier model. During the testing period, the classification assesses whether a new text is spam or not. Lastly, data classification for various machine learning techniques is carried out and the effectiveness of each machine learning algorithm has been evaluated so as to select the ideal methodology for the proposed technique. Processing is the primary move in the modification of unstructured data into structured one as characters are continually replaced for terms in text messages. To remove stop words from the SMS texts, the stop word list extractor utilized in this research work illustrates the frequency of terms in SMS messages while on the other hand, the most familiar terms that are used in spam text messages pertain to the noun and predicate categories. Likewise, pronouns, propositions, and a mixture of many other stop words stand over the top terms in spam SMS messages.

9.4 ALGORITHMS

9.4.1 KNN Algorithm

An easy and broadly used machine learning algorithm, K-Nearest Neighbors (KNN) for both the classification and regression tasks. This algorithm is non-parametric and instance-based. Here's a brief overview of how KNN works:

Training Phase: The algorithm basically stashes the whole training dataset in this phase of KNN.

Prediction/Classification Phase: To make a prediction for a new data point, this algorithm locates the k training instances that are nearest to the new point in the feature space. The "closeness" is typically measured by a distance metric, for instance, Euclidean distance. For classification, this algorithm allocates the most common class label between the k neighbors to the new data point. For regression, the algorithm assigns the target values average of the k neighbors to the new data point.

Key characteristics of KNN:

- K is a hyperparameter representing the number of neighbors to consider.
- It is a lazy learner because it doesn't build a model during training; instead, it memorizes the training data.
- KNN is sensitive to the option of distance metric as well as to the value of k.
- KNN is a straightforward algorithm and can be effective in many cases, especially when the decision boundary is not highly complex. However, it may suffer from the curse of dimensionality in high-dimensional spaces. Additionally, it requires storing the entire dataset in memory, making it memory-intensive for large datasets.

9.4.2 Naive Bayes Algorithms

It is a family of model-based machine-learning algorithms built on Bayes' theorem. Despite its simplicity, Naive Bayes is often surprisingly effective, especially for text classification and other tasks with high-dimensional data. The "naive" assumption in Naive Bayes is that features are hypothetically separate owing to the class label that simplifies the computation of probabilities.

Here are some common variants of this algorithm:

- **Gaussian Naive Bayes**: supposes that features follow a Gaussian (normal) distribution. It is suitable with respect to continuous data.
- **Multinomial Naive Bayes**: appropriate for distinct data, often utilized in text classification with word counts or term frequency features.
- **Bernoulli Naive Bayes**: suitable for binary data, often used in text classification with binary feature vectors (presence or absence of words).

Key characteristics of Naive Bayes:

- **Simple and Fast**: Naive Bayes is computationally efficient and simple to implement.
- **Assumption of Independence**: The "naive" assumption of feature independence can be limiting but often works well in practice.
- **Good for Text Data**: It is generally used for text classification tasks like spam detection and also sentiment analysis.

While Naive Bayes may not capture complex relationships in the data as well as more advanced models, its simplicity and speed make it a popular choice, especially in scenarios where interpretability and efficiency are crucial.

9.4.3 Support Vector Machines

A powerful structured machine learning algorithm, Support Vector Machines (SVM) is utilized in both classification and regression tasks. It is especially effective in high-dimensional spaces and is well suitable for scenarios where clear decision boundaries are needed. The algorithm operates by detecting the hyperplane which finely divides the data points into various classes.

Key components and concepts associated with SVM include:

➢ *Hyperplane*:
  - A hyperplane is a line that separates two classes in a two-dimensional plane. In higher-dimensional spaces, it becomes a plane.
  - The purpose of SVM locating the hyperplane that maximally divides the data points pertaining to different classes.

➢ *Support Vectors*:
  - These are the data points which are nearest to the decision boundary (hyperplane).
  - These vectors contribute a major role in expressing the optimal hyperplane.

➢ *Margin*:
- Margin is the distance connecting the hyperplane and the nearest data point of whichever of class.
- The SVM aims at maximizing this margin, providing a wider separation between classes and improving generalization to new data.

➢ *Kernel Trick*:
- The SVM algorithm is capable of handling the non-linear decision boundaries proficiently by converting the input features directly into a higher-dimensional space. This is done with the help of a kernel function.
- The common kernel functions comprise polynomial, linear, and radial basis function (RBF or Gaussian) kernels.

➢ *C Parameter*:
- The C parameter in SVM controls the trade-off between achieving a smooth decision boundary and classifying training points correctly.
- A smaller C encourages a larger margin but may misclassify some points, while a larger C aims for correct classification at the expense of a narrower margin.

One of the extensively used algorithms in a variety of applications such as image classification, text categorization, and also bioinformatics is SVM. Its effectiveness in high-dimensional spaces and the capacity to tackle nonlinear relationships make it a versatile and popular algorithm in the machine-learning community.

9.4.4 Suggested Method

We utilize Weighted Voting Classification which is a kind of ensemble learning method where the predictions related to multiple base classifiers are combined using weights. In this approach, each base classifier is assigned a weight, and these weights influence the final decision of the ensemble. The weights can be adjusted based on the performance or confidence level of individual classifiers.

9.5 PERFORMANCE ANALYSIS

For our research, we have used a dataset that contains genuine messages and fraudulent messages. First, all the null values from the dataset were removed and after that labeling of data into genuine and fraudulent was done. A total of 10,000 such labeled datasets were taken of which 8,000 were used for training and 2,000 for testing purposes.

The performance of the suggested fraud identification method is carried out by employing a sequence of testing. A feature extraction method has been chosen first over the behavior of spam and non-spam messages after which this detail would be separated out of the data in order to generate the category pattern. Various classification algorithms are used for obtaining the performance of the system. To tackle this problem, an in-built solution has been provided. At the time of split discovery, the missing data is overlooked and then set it to the section that minimizes the loss greatly. In case, if there is no need to use the choice and rather employ an alternate approach, convert the parameter to; ahead of beginning the category supposition, we ought to assess the terms in order to detect common words in fraud messages. The details that hold the info having a limited amount of probable worth have to be found out. Usually, the machine learning algorithms function instantly with categorical features and to utilize those algorithms, subsections have to be changed first into numerals prior to the application of the deep learning model upon them. The two methods that can be utilized are one-hot-encoder and values updating. For converting content variables into statuary variables, pandas have been utilized in getting statuary values. To deal with the missing values, Imputation and erasure are used. Prior to the construction of the machine learning technique, feature scaling is considered as the major important phase while compiling data in machine learning algorithms. The dataset oftentimes contains factors having a wide range of orders of magnitude, units, and ranges as well. Contrarily, utilizing the features that are unstructured causes problems in most machine learning systems. For machine learning techniques that utilize gradient descent like an optimization procedure, for instance, classification model and neural networks, the dataset needs to be increased. Anytime we make use of proximity algorithms SVM that are extremely influenced by the set of advantages, making decisions would be quite complex. This is because they utilize data point distances to estimate their similarity. There is no need to bother about its feature scalability. On tree-based algorithms, the size of the features has only minimal effect. Table 9.1 displays the analogy of different learning algorithms with respect to accuracy, precision, and recall; further, this observation is depicted in the form of a bar chart in Figure 9.1.

It is observed that the accuracy of SVM, KNN, decision tree, and Naive Bayes algorithms is 95.2%, 96.1%, 97.2%, and 98.5%, respectively. The accuracy of n the suggested *Weighted Voting Classification* base learning algorithm is more than 99% as displayed in the pictorial analogy with respect to accuracy relating to various techniques in Figure 9.1. Further the suggested technique also outperforms in terms of precision and recall.

TABLE 9.1

Performance Comparison of the Learning Model

Algorithms	Accuracy	Precision	Recall
KNN	96.1	96.0	96.1
Naive Bayes	98.5	98.4	98.5
SVM	95.2	95.1	95.2
Decision tree	97.3	97.2	97.3
Suggested	99.2	99.1	99.2

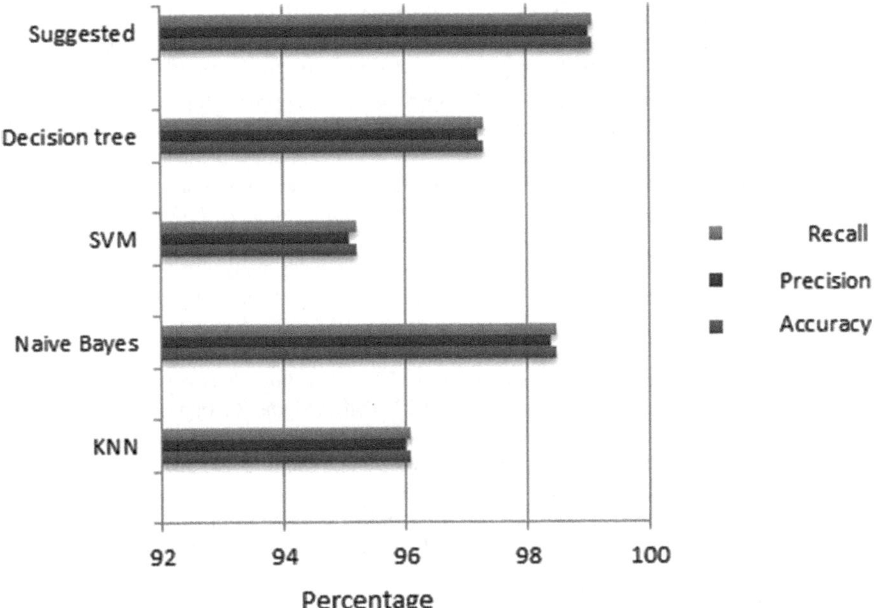

FIGURE 9.1 Bar chart of the performance comparison.

9.6 CONCLUSION

Because of the rise in the usage of text messaging every single day, the issue of SMS spam is getting more extensive. In the past few years, filtering a fraudulent message has become a major cause of concern. For our research work here, five distinct machine learning algorithms, namely decision tree, KNN algorithm, SVM algorithm, suggested *Weighted Voting Classification*, and Naive Bayes where ten features from the dataset have been utilized. On hand, one can find 10,000 labeled data inside of the collection, which includes 8,000 training and 2,000 testing class. Having an efficiency of 99.2%, the suggested learning algorithm outdoes all the other classification algorithms.

REFERENCES

Chan PPK, Yang C, Yeung DS, Wing W. Y. Ng. (2015) Spam filtering for short messages in adversarial environment. *Neurocomputing* 155:167–176. https://doi.org/10.1016/j.neucom.2014.12.034

Delany SJ, Buckley M, Greene D (2012) SMS spam filtering: methods and data. *Expert Syst. Appl* 39:9899–9908. https://doi.org/10.1016/j.eswa.2012.02.053

El-Alfy ESM, Al Hasan AA (2016) Spam filtering framework for multimodal mobile communication based on dendritic cell algorithm. *Future Gen Comput. Syst* 64:98–107. https://doi.org/10.1016/j.future.2016.02.018

Ji Z, Ma Y, Li S (2017) SVM based telecom fraud behavior identification method. *Computer Engineering & Software* 38:104–109.

Jialin M, Zhang Y, Liu J, Yu K, Wang X (2016) Intelligent SMS spam filtering using topic model. In: *International conference on intelligent networking and collaborative systems (INCoS)*, pp. 380–383. IEEE. https://doi.org/10.1109/INCoS.2016.47

Jun LG, Nazir S, Khan HU, Ul Haq A (2020) Spam detection approach for secure mobile message communication using machine learning algorithms. *Secur Commun Netw* 2020:6. https://doi.org/10.1155/2020/8873639

Li R, Zhang Y, Tuo Y, Chang P (2018) A novel method for detecting telecom fraud user. In *2018 3rd International Conference on Information Systems Engineering (ICISE)*, Shanghai, China.

Mianning H, Li X, Li M, Zhu R, Si B (2023) A framework for analyzing fraud risk warning and interference effects by fusing multivariate heterogeneous data: A bayesian belief network. *Entropy 2023* 25(6):892. https://doi.org/10.3390/e25060892

Nuruzzaman MT, Lee C, Abdullah M, Choi D (2012) Simple SMS spam filtering on independent mobile phone. *Secur Commun Netw* 5:1209–1220. https://doi.org/10.1002/sec.577

Nuthakki R, Aameen A, Kumar N, Mishra SK (2023) Traffic Signal Recognition System Using Deep Learning. *2023 International Conference on Sustainable Emerging Innovations in Engineering and Technology (ICSEIET)*, Ghaziabad, India, 2023, pp. 636–639. https://doi.org/10.1109/ICSEIET58677.2023.10303609

Nuthakki R, Masanta P, Yukta TN. (2021) Speech Enhancement based on Deep Convolutional Neural Network. *2021 Fifth International Conference on I-SMAC (IoT in Social, Mobile, Analytics and Cloud) (I-SMAC)*, 2021, pp. 1–6. https://doi.org/10.1109/I-SMAC52330.2021.9640736

Nuthakki R, Masanta P, Yukta TN (2022). *A Literature Survey on Speech Enhancement Based on Deep Neural Network Technique. Lecture Notes in Electrical Engineering*, vol. 828. Springer, Singapore. https://doi.org/10.1007/978-981-16-7985-8_2

Sahin M, Francillon A, Gupta P, Ahamad M (2017) SoK: Fraud in telephony networks. In *2017 IEEE European Symposium on Security and Privacy (EuroS&P)*, pp. 235–250, Paris, France, 2017.

Sanver M, Karahoca A, "Fraud detection using an adaptive neuro-fuzzy inference system in mobile telecommunication networks," *Journal of Multiple-Valued Logic and Soft Computing* 15(2):155–179, 2016.

Uysal AK, Gunal S, Ergin S, Gunal ES (2012) A novel framework for SMS spam filtering. In: *International Symposium on Innovations in Intelligent Systems and Applications (INISTA)*, pp. 1–4. IEEE. https://doi.org/10.1109/INISTA.2012.6246947

Xing J, Yu M, Wang S, Zhang Y, Ding Y (2020) Automated fraudulent phone call recognition through deep learning. *Wireless Communications and Mobile Computing* 2020. https://doi.org/10.1155/2020/8853468

Xu Q, Xiang EW, Yang Q, Du J, Zhong J (2012) SMS spam detection using non-content features. *IEEE Intell Syst* 27(6):44–51

Yadav K, Kumaraguru P, Goyal A, Gupta A, Naik V (2011) SMS assassin: crowdsourcing driven mobile-based system for SMS spam filtering. In: *12th Workshop on Mobile Computing Systems and Applications*, pp. 1–6. ACM. https://doi.org/10.1145/2184489.2184491

Zhou G, Chen G, Zhou Y (2015) User behavior in telecommunication fraud based on CDR analysis. In *Information Security and Communications Privacy*, pp. 114–118. The 30th Research Institute of China Electronics Technology Group Corporation, 2015.

10 Artificial Intelligence of Things (AIoT)-Based Telehealth System Using Healthy Pi

Arif Hussain, Abid Iqbal, Muhammad Abeer Irfan, Irfan Ahmed, and Amaad Khalil
University of Engineering & Technology Peshawar, Peshawar, Pakistan

10.1 INTRODUCTION: BACKGROUND AND DRIVING FORCES

The world's population has increased enormously in the last few decades, increasing health-related problems (Raleigh 1999). According to the world death clock, 56 million people die yearly. The primary cause of the deaths is health-related issues. In developing countries, health conditions worsen, and the number of deaths increases. There has always been a shortage of patient beds in hospitals in developing countries. A survey was conducted by World Health Organization (WHO) in 2017 for beds available per 1,000 patients, which came out to be 0.63. The report revealed that almost half of the patients could not get hospital beds. The medical staff, medicines, equipment, and other facilities and infection safety systems are also scarce. The need of the hour is to have a telemedicine system where health specialists can monitor patients despite the considerable distance separating them. The number of patients visiting hospitals can be reduced by using the modern techniques of telemedicine-based on the Internet of Things (IoT) and Artificial Intelligence (AI) (Elhennawy et al. 2021).

The IoT is expanding at unfathomable rates, bringing about technological and cultural changes for businesses, organizations, and industries as they transform a more intelligent model that considers social and environmental responsibility. Implementing IoT in health-monitoring systems has significant advantages, including control over medical devices and medications, management of medical information, telemedicine, and mobile medical care, as well as implementation strategies and methodologies that include a rehabilitation system (Mohammed et al. 2020).

The world has changed with the modern and emerging technologies of AI and IoT. The AI and IoT are called Artificial Intelligence of Things (AIoT) and have brought a great revolution in almost every field of life (Rouhiainen 2018). One of these applications is in healthcare to implement a telehealth system to monitor a patient's vital

DOI: 10.1201/9781032644509-10

health parameters to the doctor for remote diagnosing. The system can become a doctor predict abnormality in the patient's health status using AI (Mintz and Brodie 2019).

The IoT is a system that incorporates many physical objects, subsystems, and sensors, which are interrelated and connected to the internet. These devices communicate information with each other wirelessly without the intervention of any human. AI, in simple words, is the mimicry of human/natural intelligence using both technologies at once and is known as Artificial Intelligence of Things (González García et al. 2019). Applying IoT in the health field enhances the quality of service through the acquisition of diagnoses and more precise analysis. Using IoT applications for remote diagnostics saves time and money by lowering the number of periodic patient reviews of the hospital. Also, the researchers can quickly receive the patient's data from a well-organized and adequately structured IoT (Kadhim et al. 2020).

One of the many uses of IoT is activity recognition. It enables individuals to monitor their health and maintain it independently. Additionally, it aids older people who need continual supervision from falling or slipping due to no fault of their own. The same is valid for keeping an eye out for heart disease, blood pressure monitoring, sleep monitoring, Alzheimer's disease monitoring etc., and quick action in case of any problems (Mamdiwar et al. 2021).

IoT is important for tracking a person's health and COVID-19 diagnosis. Chakraborty et al. reviewed the role of machine learning (ML) for reliable COVID-19 detection and diagnosis from X-ray and CT images. ML can be employed for COVID-19 segmentation and prediction. Moreover, ML can efficiently enhance drug discovery and lower clinical failure rates (Chakraborty and Abougreen 2021).

Sarkar et al. proposed a fast and reliable neural network-based security system based on whale optimization-based neural synchronization for key exchange protocol development. A special neural network structure called Double Layer Tree Parity Machine (DLTPM) is proposed for neural synchronization. The DLTPM model used a whale algorithm-optimized weight vector that was faster and more secure (Sarkar et al. 2021).

AI is frequently used in healthcare 4.0 for quick and reliable outcomes. Kishor et al. created a healthcare model to assist doctors in making an early disease diagnosis. Nine fatal diseases, including heart disease, diabetes, breast cancer, hepatitis, liver disorder, dermatology, surgery data, thyroid, and spec heart, were predicted using seven machine learning classification algorithms by Kishor et al. They used accuracy, sensitivity, specificity, and area under the curve as the performance indicators of their model. They observed that the RF classifier was the best, with a maximum accuracy of 97.62%, a sensitivity of 99.67%, a specificity of 97.81%, and an AUC of 99.32% (Kishor and Chakraborty 2022).

Islam et al. proposed an intelligent healthcare system for the IoT that can continuously track a patient's vital signs. They employed five sensors, i.e., the heartbeat, body temperature, room temperature, CO, and CO_2 sensor, to collect data from the hospital environment and send it over IoT for medical staff to assess and evaluate the patient's current state (Islam and Rahaman 2020). An Internet of Things-based wearable, portable, low-power, real-time remote biosignals monitoring system was implemented. In that project, a smartphone app was used as an IoT platform to remotely monitor patients' body temperatures, heart rates, SpO_2, and live ECG signals. A microcontroller-based device was used to measure and process the signals (Arduino). The electrocardiogram (ECG) signal was successfully transmitted to a particular smart mobile device (Al-Sheikh and Ameen 2020).

Machine learning is the branch of artificial intelligence that plays a vital role in diagnosing many diseases like cancer. Since medical data has been converted to digital form, machine learning has become crucial in detecting numerous diseases, including cancer. Researchers have examined various machine learning techniques in the healthcare industry in the past ten years. These methods function as (a) abnormality segmentation or detection and (b) classifying the segmented abnormality as malignant. Kaur et al. used machine learning to evaluate prediction systems for liver disorders, heart disease, breast cancer, diabetes, thyroid, dermatology, and surgical data. They used and compared the machine learning techniques, including K-NN, Support Vector Machine, Decision Trees, Random Forest, and MLP. They observed that on the Dermatology dataset, the Random Forest machine learning algorithm has a maximum accuracy of 97.26% (Kaur et al. 2019).

Souri et al. proposed an IoT-based student healthcare monitoring model that continually monitors students' vital signs and uses smart healthcare technologies to identify biological and behavioral changes. They gathered the crucial information using IoT devices, while data analysis was done using machine learning techniques to identify the likely dangers of students' physiological and behavioral changes. The experimental findings show that the suggested model is adequate and accurate for determining the state of the students, with the highest accuracy of 99.1% for the Support Vector Machine algorithm (Souri et al. 2020). Kondaka et al. introduced a new algorithm called iCloud-Assisted Intensive Deep Learning (iCAIDL), which supports healthcare providers and patients by applying machine learning techniques and an intelligent cloud system. The proposed algorithm was derived from deep learning norms as its foundation. They designed a Smart Medical Gadget to collect patient's health records and heart rate, pressure level, and blood flow. They claimed that their designed system provides a robustness guarantee for health data, as data loss or corruption is frequently a problem while maintaining medical records (Kondaka et al. 2021). Godi et al. propose the E-Healthcare Monitoring System (EHMS) based on an IoT application framework that integrates machine learning methods to construct a sophisticated automation system. They demonstrated their model via a raw diabetic data set from online sources (Godi et al. 2020). Raza et al. designed a system to track and forecast the development of Parkinson's disease (PD). The indoor IoT framework helped collect auditory samples from PD patients and other basic sensory data. They employed machine learning techniques to predict the progression of PD using auditory input (Raza et al. 2020).

Gondalia et al. designed a system to track the location and monitor the health of the soldiers in real time on the battlefield. They utilized the wireless body area sensor networks (WBASNs) for health monitoring, ZigBee for wireless transmission, and GPS for position location tracking. Furthermore, they utilized the LoRa WAN network infrastructure between the squadron leader and the control unit in high-altitude warzones where cellular network coverage is either absent or does not allow data transmission. They uploaded the collected data to the cloud and utilized the K-Means Clustering algorithm for further data analysis and predictions (Gondalia et al. 2018).

The people, especially the old, paralyzed, handicapped, and those living far from rural areas, cannot visit hospitals in cities/towns for daily checkups. They can monitor their vital health parameters using AI. Moreover, they can have a live/online meeting with their health specialists, doctors, or caretakers from their homes and villages with health parameters, information, and prescription shared between both sides in real time using the technologies of IoT and cloud computing (Cao et al. 2021). Naik et al. (2019) monitor somebody's health parameters using Raspberry Pi v3. They used a temperature sensor LM35, pulse rate sensor, ECG, and blood pressure connected to the Raspberry Pi. The patient data was uploaded on the Thing Speak cloud through these sensors using Raspberry Pi and shared with doctors. Ganesh (2019) proposed a health monitoring system that uses a Raspberry Pi v3 model B and a blood pressure sensor. The data from the patient through Raspberry Pi is monitored and stored in the cloud. The patient can share the data with the doctor; if the data exceeds a certain threshold, a message will be sent to the doctor.

In most previous works, the healthcare systems were designed to measure the data from the human body using various sensors and upload it to the cloud using IOT. In some scenarios, they employed machine learning and deep learning techniques to classify chronic diseases and abnormal conditions from patient data.

The novelty of the system discussed in this chapter is that we proposed a low-cost and easy-to-use health monitoring system that collects vital data from the human body, detects an abnormality, and informs the paramedics/caretakers. Our proposed telehealth system consists of sensors, Healthy Pi, and Raspberry Pi that collect the patient's data from the human body and upload it to the cloud via Node-Red server using the IoT. The design system applies machine learning algorithms to the data that predicts the abnormality in human health status and establishes an online meeting between the patient and the doctor incorporating the patient's health sensor data.

10.2 SYSTEM ARCHITECTURE

The system architecture of our proposed system is explained with the help of the flow diagram, as shown in Figure 10.1. The system starts with circuit initialization. It proceeds with collecting patient data through sensors if the power supply and other connections are suitable. The hardware incorporates temperature, pulse oximetry, respiration, and heart rate sensors. The data taken is filtered, processed, displayed, and sent to the cloud. The cloud's already deployed machine learning algorithm is applied to the latest coming sensor data for every 5 seconds of timestamps. If the data are normal, the algorithm ends with displaying the message "No worry, your health conditions are good." If the data has abnormalities, the system proceeds with an alert message "Your health conditions are not looking good. Do you want to send an alert mail to your doctor?" If the patient agrees, send an alert mail to the authorized doctor or caretaker; otherwise, the system will end.

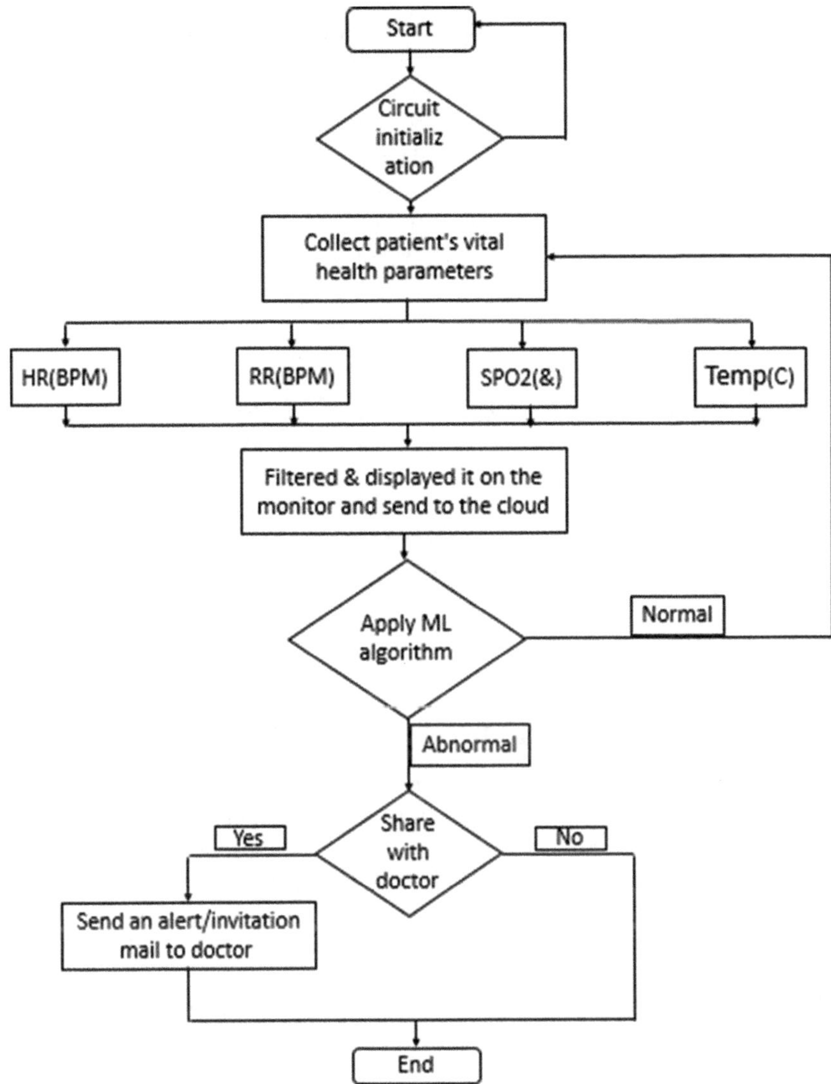

FIGURE 10.1 Flow diagram of our designed system.

The system block diagram is shown in Figure 10.2. Healthy Pi v3 HAT circuit board is used to assemble sensor data, filter it, and send it to the Raspberry Pi. The Raspberry Pi is used as the main central processing unit. LCD, pi camera, keyboard, mouse, and other modules are connected to it. The Node-RED bridges the cloud and the local Raspberry Pi dashboard. The Raspberry Pi displayed, processed the data, and sent it to the IBM IoT platform. The IBM dashboard then decides the normal and abnormal data by applying it to the machine learning algorithm. The data can be seen by anyone around the globe having an invitation link or authorization access.

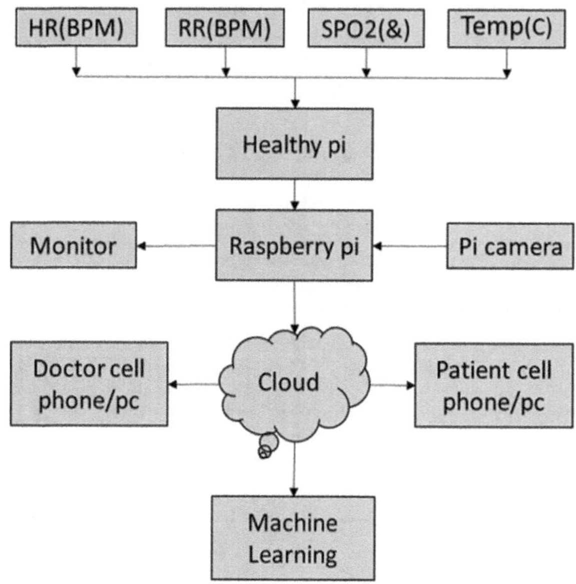

FIGURE 10.2 Block diagram of our designed system.

10.3 THE ARCHITECTURE OF THE SMART HEALTHCARE SYSTEM

The designed smart healthcare system can decide and alert the patient's observed conditions, such as body temperature, heartbeats, and pulse rate. The proposed healthcare system uses only one sensor at a time to save energy, increasing energy efficiency. The system's algorithmic study will manage the sensors' working period and so may control the cost of the system. The study in reference addresses the use of remote monitoring of the patient's health for the necessary treatments via the doctors in the hospital.

The embedded internal and external sensors, IoT server, communication channels, and cloud storage are used to monitor and administer the healthcare system of patients proposed in this study. Figure 10.3 depicts the entire architecture of the proposed system. The essential principle that ensures the efficiency of telemedicine in rural regions is the use of sensors and a supported decision system in telemedicine.

10.4 IMPLEMENTATION AND CLASSIFICATION OF THE SMART HEALTHCARE SYSTEM

The proposed system is designed to implement a smart telehealth device for domestic use and in remote clinics with the help of AI and IoT, collectively called the Internet of Things. For clinics, the designed device will collect the patient's data, such as body temperature, heart rate, respiration, and pulse oximetry, as inputs and will transfer this data to the doctors in the hospital. In connection with the data analyzed, the doctor will monitor the patient's condition and direct the remote area clinic crew

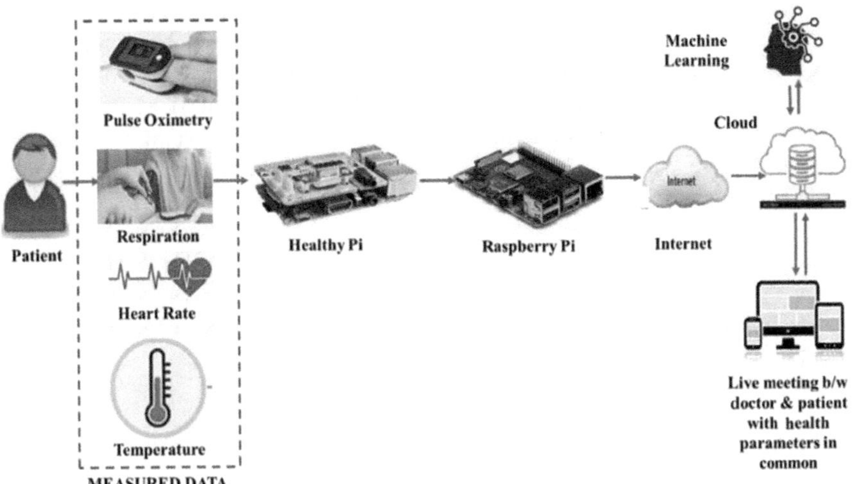

FIGURE 10.3 Smart healthcare system architecture.

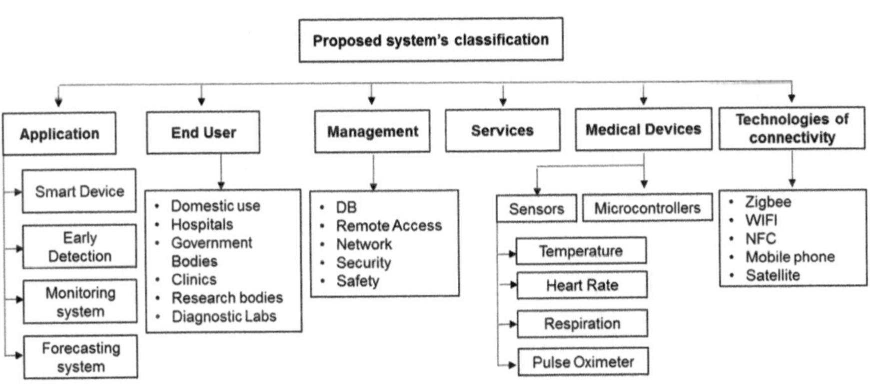

FIGURE 10.4 Classification of the smart healthcare system.

to the necessary steps for the best patient treatment. Domestic use is straightforward. Patients can use it for daily checkups, and if the smart device detects any abnormality, they can share their vital parameters with authorized doctors and have a meeting. Figure 10.4 shows the classification of the proposed smart healthcare system.

The body temperature, pulse rate, respiration, and pulse oximetry are the four sensors that make up the system. These sensors are attached to the Healthy Pi board to gather and store patient's data, with data transmission handled via networking equipment. The acquired data demonstrates analytics that may provide decision-making capabilities, and Random Forest classifier systems are used in the decision-making process. As a result, the doctor may be able to deliver consultations to patients in far-flung locations.

A Random Forest is an ensemble technique that uses several decision trees and a technique called Bootstrap and Aggregation, sometimes called bagging, to solve regression and classification problems. Instead, depending on individual decision trees, the main idea is to aggregate numerous decision trees to determine the outcome. As a fundamental learning model, Random Forest uses several decision trees. Row and feature sampling are done randomly from the dataset, resulting in sample datasets for each model.

10.5 DESIGN METHODOLOGY

The design section contains the hardware and software components used in our model.

10.5.1 Hardware Components

The Healthy Pi is employed, which is a full-featured, open-source vital sign monitor. It acts as a HAT (hardware attached on top) with Raspberry Pi, which makes it a vital sign monitoring system by employing it as a computer and display platform. The hardware contains the four basic human vital sign sensors with it. The Healthy Pi has been made by modifying the Arduino zero. The central chip Atmel's SAMD21 MCU is common in Healthy Pi and Arduino zero. Healthy Pi is a valuable prototype for healthcare system designers. SpO2 Probe is a sensor device used to measure the oxygen saturation level in the blood vessels. It uses optical properties to measure oxygen contents in the blood. The MAX30205 is our model's high-accuracy digital human body temperature sensor. It communicates directly through an I2C interface. It incorporates a sigma-delta high-resolution analog-to-digital converter (ADC), which converts analog measures of temperature values to digital form (Kaur et al. 2019). Analog front-end IC TI ADS1292R measures both ECG and respiration values. It uses a three-electrode cable. The two electrodes are for ECG, and one Driven Right Leg (DRL) electrode is for noise reduction. The IC also measures respiration using "impedance pneumography," a measure of the change in chest impedance.

10.5.2 Software Components

The Healthy Pi uses Arduino Integrated Development Environment (IDE), an application whose functions are written in C and C+. For the Healthy Pi SAMD cortex, MO+ Arduino zero board was selected in the settings. Python is a high-level, widely used general purpose programming language. Because of the quality and simplicity of Python language, we used it for artificial intelligence applications.

Node-RED is a flow-based software incorporating visual programming that connects APIs, hardware devices, and online services based on the Internet of Things. IBM Watson IoT platform is a cloud server mainly used for the Internet of Things. It allows users to connect, analyze, predict, and manage IoT data. Its portfolio is designed to make it easy for users to handle data from various sources and get more value from AI. The Node-RED is a product of IBM as well and its interface is shown in Figure 10.5; hence, it is so compatible with the Watson IoT platform. It provides excellent visuals for sensor data and is a good platform for real-time monitoring.

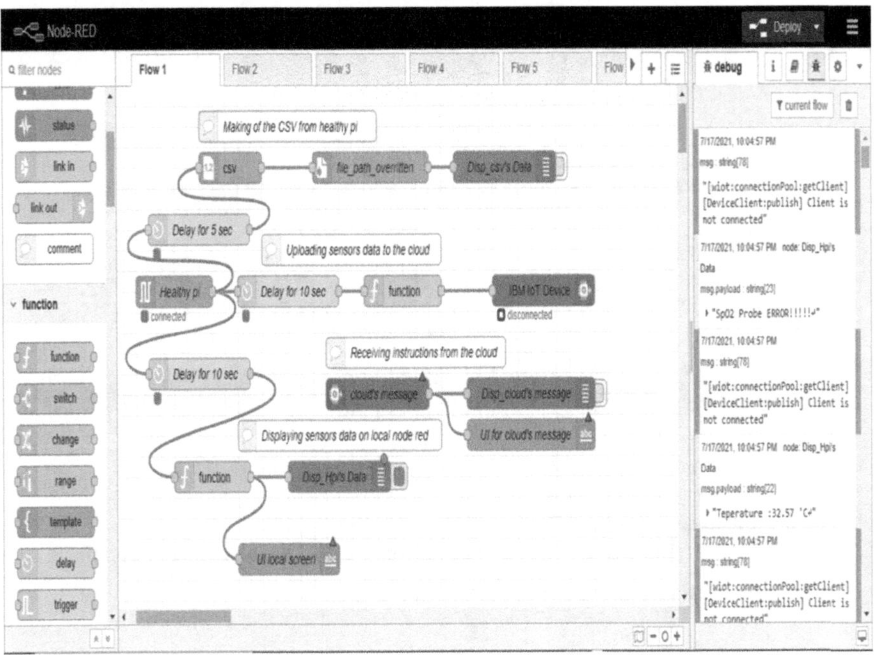

FIGURE 10.5 Node-RED interface and connections for our designed system.

10.6 RESULT AND DISCUSSION

The proposed system is both software- and hardware-based smart healthcare system, which is comprised of mainly two emerging technologies:

- The Internet of Things (IoT) and
- Artificial Intelligence (AI).

The data collected from the patient using human vital sign sensors (temperature, heart rate, respiration, and pulse oximetry with Healthy Pi) is successfully transmitted to the IBM cloud using the Node-Red server, as shown in Figure 10.5. The transmitted data is successfully monitored in the cloud for remote access. After the doctor analyzes and diagnoses, the doctor's prescription is sent back to the patient through the same channel and displayed to the patient. Hence communication is built between patient and doctor, as shown in Figure 10.6, which was far from the Internet of Things part.

Meanwhile, the device was trained using supervised machine learning. It could predict the patient's health status abnormality by comparing the data with the already trained and deployed machine learning model on the cloud using Random Forest regressor and flask, respectively. The predicted results were shown to the patient. The data were preprocessed using label encoding, one hot encoding, standard scaling, etc. The dataset used in our model is a public dataset from Kaggle containing vital human signs predicting binary output as normal or abnormal (Kagle Reference). The input variables have a specific range that predicts an output as normal or abnormal.

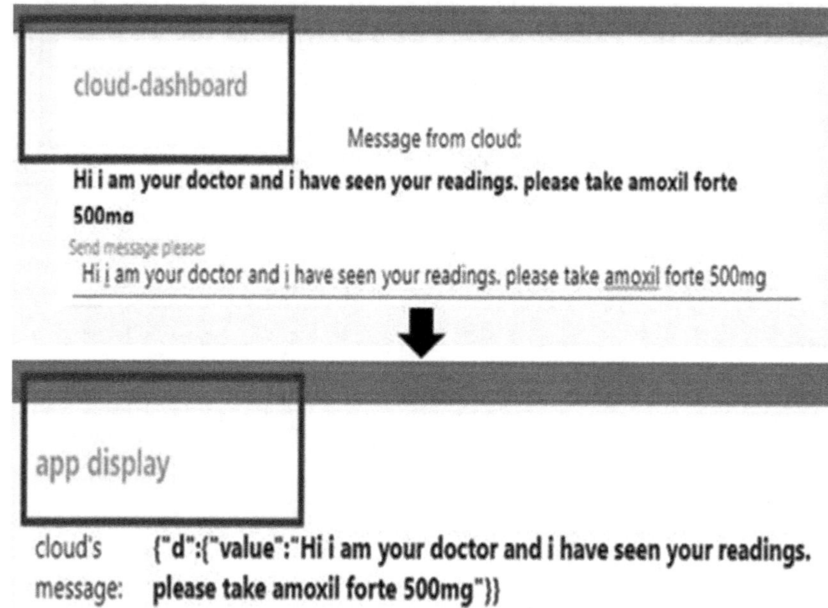

FIGURE 10.6 Communication protocol between patient and doctor

TABLE 10.1
Input Sensors Values Classification

Type of Sensor	Range	Normal/Abnormal
Temperature	T ≤ 37C	Normal
	T > 37C	Abnormal
Heart Rate	60bpm ≤ HR ≤ 100bpm	Normal
	60bpm > HR > 100bpm	Abnormal
Respiration	12bpm ≤ R ≤ 16bpm	Normal
	12bpm > R > 16bpm	Abnormal
Oxygen Saturation	OS ≥ 95%	Normal
	OS < 95%	Abnormal

The classification of the used sensor's values as normal or abnormal is given in Table 10.1. The algorithm is set such that if any sensor's value is abnormal, the system will display an abnormal health condition.

The best accuracy of 97% was achieved using Random Forest Classifier, the best-supervised machine learning algorithm involving many input variables predicting the binary output. The classification report, i.e., our model's precision, recall, f1 score, and accuracy under different algorithms, is given in Table 10.2. The confusion matrix is shown in Figure 10.7, which shows that the model is trained relatively better under a random forest classifier.

TABLE 10.2
Classification Report

Algorithm	Precision (%)	Recall (%)	F1 Score (%)	Accuracy (%)
Random Forest	95	98	96	97
Decision Tree	95	97	97	95
Logistic Regression	93	93	93	94
KNN	95	98	96	96
SVM	94	98	95	96

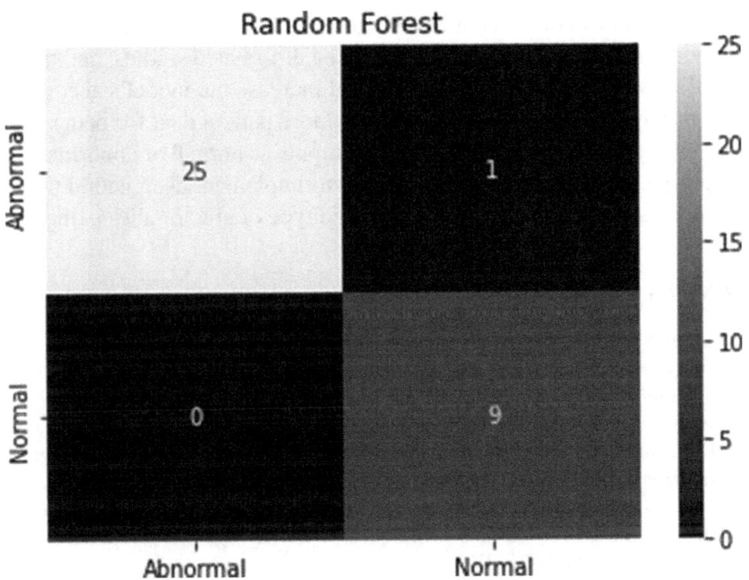

FIGURE 10.7 Confusion matrix for Random Forest.

We tested our proposed system on 101 people, and the results were entirely satisfactory. Our system detected 11 real-life abnormal respiration, blood saturation, and fever patients with an accuracy of 97%. We appended the collected data from patients to the dataset that was used for training our system for better accuracy.

Compared to other systems, the novelty of our proposed system is a smart health device that collects and processes data via machine learning algorithms to predict abnormalities that will reduce the unnecessary burden on hospitals. It also includes sending prescriptions back to patients through the same channel, making it a complete telemedicine system (Senthamilarasi et al. 2018; Ghosh, Halder, and Hossain 2016; Neela n.d.; Gupta et al. 2019).

10.7 CONCLUSION AND FUTURE WORK

In this chapter, we presented a telehealth device that takes human vital health parameters and transmits them to the IBM cloud for remote monitoring. The proposed device can predict abnormality in the patient's vital parameters data by comparing it to the already trained and deployed model using machine learning. The doctors/caretakers can analyze the patient's incoming readings in real time and send their prescriptions back to the patient. They also can have a live video meeting using a pi camera with incoming signals from the patient in common. The designed device is best for daily domestic health checkups and clinics.

The proposed approach has been generalized so far. It can be customized for critical conditions like operation theatre, intensive care unit patients, newborn babies, and more complex patients. A particular database could be created online or offline to store patients' old medical records and age and analyze vital signals. This database would provide helpful information for future health-related studies, and training the deep neural network with this database would increase the model's accuracy. In the future, we propose using more sensors to get more patient data for better diagnosis. So far, we have used regression, i.e., binary outputs as normal or abnormal, as this is just a prototype. The system can be modified to implement deep neural networks to obtain more classes of outputs, i.e., the specific type of abnormality in the patient.

REFERENCES

Al-Sheikh, Mustafa A, and Ibrahim A Ameen. 2020. "Design of mobile healthcare monitoring system using IoT technology and cloud computing." *IOP Conference Series: Materials Science and Engineering*, 881, 012113–012131.

Cao, Shihua, Xin Lin, Keyong Hu, Lidong Wang, Wenjuan Li, Mengxin Wang, and Yuchao Le. 2021. "Cloud computing-based medical health monitoring IoT system design." *Mobile Information Systems* 2021, 1–12.

Chakraborty, Chinmay, and Arij Technology Abougreen. 2021. "Intelligent internet of things and advanced machine learning techniques for Covid-19." *EAI Endorsed Transactions on Pervasive Health* 7 (26), E1.

Elhennawy, Amr, Fateh Almohammed Alsalem, Salah Bahri, and Noor Alarfaj. 2021. "Telemedicine versus Physical Examination in Patients' Assessment during COVID-19 Pandemic: The Dubai Experience." *Dubai Medical Journal* 4 (2):175–180.

Ganesh, EN. 2019. "Health monitoring system using raspberry Pi and IoT." *Oriental J Comput Sci Technol* 12 (1):08–13.

Ghosh, Ananda Mohon, Debashish Halder, and SK Alamgir Hossain. 2016. "Remote health monitoring system through IoT." *2016 5th International Conference on Informatics, Electronics and Vision (ICIEV)*.

Godi, Brahmaji, Sangeeta Viswanadham, Appala Srinuvasu Muttipati, Om Prakash Samantray, and Sasi Rekha Gadiraju. 2020. "E-healthcare monitoring system using IoT with machine learning approaches." *2020 International Conference on Computer Science, Engineering and Applications (ICCSEA)*.

Gondalia, Aashay, Dhruv Dixit, Shubham Parashar, Vijayanand Raghava, Animesh Sengupta, and Vergin Raja Sarobin. 2018. "IoT-based healthcare monitoring system for war soldiers using machine learning." *Procedia Computer Science* 133:1005–1013.

González García, Cristian, Edward Rolando Núñez Valdéz, Vicente García Díaz, Begoña Cristina Pelayo García-Bustelo, and Juan Manuel Cueva Lovelle. 2019. "A review of artificial intelligence in the internet of things." *International Journal Of Interactive Multimedia Artificial Intelligence* 5 (4):9–20.

Gupta, Aanshi, Shubham Yadav, Shaik Shahid, and U Venkanna. 2019. "HeartCare: IoT based heart disease prediction system." *2019 International Conference on Information Technology (ICIT).*

Islam, Md, and Ashikur Rahaman. 2020. "Development of smart healthcare monitoring system in IoT environment." *SN Computer Science* 1(3):1–11.

Kadhim, Kadhim Takleef, Ali M Alsahlany, Salim Muhsin Wadi, and Hussein T Kadhum. 2020. "An overview of patient's health status monitoring system based on Internet of Things (IoT)." *Wireless Personal Communications* 114 (3):2235–2262.

Kaur, Pavleen, Ravinder Kumar, Munish Kumar. 2019. "A healthcare monitoring system using random forest and internet of things (IoT)." *Multimedia Tools and Applications* 78 (14):19905–19916.

Kishor, Amit, and Chinmay Chakraborty. 2022. "Artificial intelligence and internet of things based healthcare 4.0 monitoring system." *Wireless Personal Communications* 127 (2):1615–1631.

Kondaka, Lakshmi Sudha, M Thenmozhi, K Vijayakumar, Rashi Kohli, and Applications. 2021. "An intensive healthcare monitoring paradigm by using IoT based machine learning strategies." *Multimedia Tools* 81(26):1–15.

Mamdiwar, Shwetank Dattatraya, Zainab Shakruwala, Utkarsh Chadha, Kathiravan Srinivasan, and Chuan-Yu Chang. 2021. "Recent advances on IoT-assisted wearable sensor systems for healthcare monitoring." *Biosensors* 11 (10):372.

Mintz, Yoav, and Ronit Brodie. 2019. "Introduction to artificial intelligence in medicine." *Minimally Invasive Therapy & Allied Technologies* 28 (2):73–81.

Mohammed, MN, SF Desyansah, S Al-Zubaidi, and E Yusuf. 2020. "An internet of things-based smart homes and healthcare monitoring and management system." *Journal of Physics: Conference Series* 1450: 012079.

Naik, Seena, and E Sudarshan. 2019. "Smart healthcare monitoring system using raspberry Pi on IoT platform." *ARPN Journal of Engineering Applied Sciences* 14 (4):872–876.

Neela, Mary Margarat Valentine. 2018. "Health monitoring system." *IC-CSOD-2018 Conference Proceedings*, 83.

Raleigh, Veena Soni. 1999. "World population and health in transition." *BMJ* 319 (7215): 981–984. doi: 10.1136/bmj.319.7215.981

Raza, Mohsin, Muhammad Awais, Nishant Singh, Muhammad Imran, and Sajjad Hussain. 2020. "Intelligent IoT framework for indoor healthcare monitoring of Parkinson's disease patient." *IEEE Journal on Selected Areas in Communications* 39 (2):593–602.

Rouhiainen, Lasse. 2018. *Artificial Intelligence: 101 things you must know today about our future*: Lasse Rouhiainen.

Sarkar, Arindam, Mohammad Zubair Khan, Moirangthem Marjit Singh, Abdulfattah Noorwali, Chinmay Chakraborty, and Subhendu Kumar Pani. 2021. "Artificial neural synchronization using nature inspired whale optimization." *IEEE Access* 9:16435–16447.

Senthamilarasi, C, J Jansi Rani, B Vidhya, and H Aritha. 2018. "A smart patient health monitoring system using IoT." *International Journal of Pure Applied Mathematics* 119 (16):59–70.

Souri, Alireza, Marwan Yassin Ghafour, Aram Mahmood Ahmed, Fatemeh Safara, Ali Yamini, and Mahdi Hoseyninezhad. 2020. "A new machine learning-based healthcare monitoring model for student's condition diagnosis in Internet of Things environment." *Soft Computing* 24 (22):17111–17121.

11 Artificial Intelligence for Social Good, Disaster Relief, Poverty Alleviation, and Environmental Sustainability

A New Era of Innovative Solution and Global Impact

V. Selvalakshmi
Velammal College of Engineering and Technology, Madurai, India

Durdana Ovais
The BSSS Institute of Advanced Studies, Bhopal, India

Shubhra Bhatia
Bhopal School of Social Sciences, Bhopal, India

S. Gayathri
Rajalakshmi School of Business, Chennai, India

11.1 INTRODUCTION: BACKGROUND AND DRIVING FORCES

The media, federal agencies, and society at large are all currently showing a great deal of interest in artificial intelligence (AI). AI has seen its share of ups and downs in public interest, from its beginnings in the 1950s through the early hopeful projections of its inventors to some recent unfavorable opinions put forth by the media (Barrat, 2023). Depending on how we shape this new technology and the questions we use to motivate young researchers, AI has the potential to be a powerful force for social good (Taddeo and Floridi, 2021). We have now entered a unique and exciting period

 DOI: 10.1201/9781032644509-11

in AI history, though, due to the consistent advancements made in basic AI research over the previous 50–60 years, the availability of immense amounts of data, and enormous gains in computing power.

It is also critical to recognize the potential complexity of interactions between humans and AI agents, as well as the growing need for laws and certification processes for AI systems that are based on ethical principles (Shneiderman, 2020; Floridi et al., 2021; de Almeida et al., 2021; Stahl, 2021). This holds true for all AI applications, but it's particularly important for those like autonomous weapons that, if left unchecked, might have disastrous consequences for humanity. The societal benefits that artificial intelligence is delivering and can deliver in the near future, as well as how our actions today can shape the future of AI, must not be overlooked. While understanding and grappling with these concerns, as well as shaping the long-term future, is a legitimate aspect of future AI research and policy-making decisions. An ever-widening array of industries are being shaped by artificial intelligence.

An evaluation of artificial intelligence's effects on social good, disaster relief, poverty elevation, and environmental sustainability is necessary given the rise of AI and its increasingly broader impact on many sectors (Shi et al., 2020). The ongoing environmental challenges and the increasing complexity of achieving sustainable development necessitate a reassessment of individuals' sustainable knowledge, attitudes, and behaviors (Ovais, 2023). Utilizing AI can assist in evaluating our current position and guiding us toward the desired sustainability goals. This evaluation will help to foster a new era of innovative solutions and global impact. It's critical to acknowledge how these problems are related to one another. Humanity has always faced this problem, which is impossible to resolve on its own since it calls for the involvement of all relevant parties. Unfortunately, no published study has evaluated the degree to which AI is now used in this manner.

We now have the power to influence how AI research develops. As a result, in order to analyze how artificial intelligence can affect social good, disaster relief, poverty elevation, and environmental sustainability, as well as the obstacles it may provide, a study of the impacts and challenges is required.

A thorough literature study and bibliometric analysis are required to map out these because a large portion of the AI expertise is now engaged in academic or industrial research. This will inspire important stakeholders in top research laboratories to further empower researchers. Here in this chapter, we present and discuss implications of how AI can either enable or inhibit the delivery for social good disaster relief poverty elevation and environmental sustainability. The study will enable researchers to take initiatives and direct their focus and efforts where appropriate and possible. The complexity of real-world challenges can in fact help to boost the understanding of existing methods and demonstrate impact where it matters the most. The study helps in shaping a strategy around how to tackle the world's most pressing challenges with some of the most powerful technological solutions available. Only together can we build a better future.

As we navigate the evolution of AI research, understanding its impacts and challenges is essential. The study, through a systematic literature review and bibliometric analysis, aims to map the current landscape of AI engaged in research. By presenting and discussing the implications of AI on social good, disaster relief, poverty alleviation, and environmental sustainability, the research enables researchers to take informed initiatives. The study contributes to shaping a strategy for addressing global challenges with powerful technological solutions, emphasizing the collaborative effort needed for building a better future.

11.2 RESEARCH QUESTION

How can artificial intelligence be effectively leveraged to address disaster relief, poverty elevation, and environmental sustainability as integral components of social good initiatives, considering the multifaceted challenges and implications associated with its application in these critical societal domains?

11.3 RESEARCH OBJECTIVES

1. To conduct a comprehensive literature review on the intersection of AI and disaster relief, poverty alleviation, and environmental sustainability, aiming to synthesize diverse theories and insights from experts in the field.
2. To explore the scattered and linear perspectives prevalent in past research, seeking a holistic understanding of the challenges and implications associated with AI in addressing complex societal issues, and thereby bridge the existing knowledge gap.
3. To critically evaluate the multifaceted nature of AI's role in social good initiatives, with a focus on disaster management, poverty reduction, and environmental conservation, in order to provide nuanced insights for informed decision-making and the development of sustainable solutions.

11.4 CONCEPTUAL FRAMEWORK

The study's conceptual framework (as shown in Figure 11.1) elucidates the AI factors that underpin social good disaster relief poverty elevation and environmental sustainability are detailed as follows:

Artificial intelligence stands as a powerful force with a profound interconnectedness to individuals, society, and the environment. In the domain of social good, AI projects actively contribute to global goals, shaping positive impacts in everyday life and fostering community engagement (Sætra, 2022). The seamless integration of AI in disaster management enhances societal resilience and individual safety, showcasing the symbiotic relationship between technology and human well-being. AI's role in poverty elevation directly influences economic landscapes, revolutionizing sectors like agriculture and finance. The interconnectedness of AI with Sustainable Development Goals (SDGs) emphasizes its pivotal role in driving societal and environmental progress (Bachmann et al., 2022; Bibri et al., 2024). Furthermore, in environmental sustainability, AI's foundational role in eco-cities showcases a harmonious alignment with societal and environmental well-being (Wilkes-Allemann et al., 2023; Bibri et al., 2024). This intricate web of interconnectedness signifies that AI

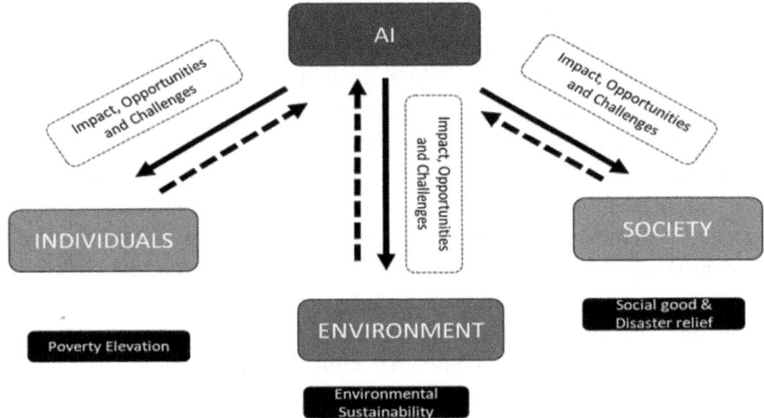

FIGURE 11.1 Artificial intelligence for social good disaster relief poverty elevation and environmental sustainability a new era of innovative solution and global impact.

Source: Developed by authours for the study.

advancements not only shape individual experiences but also have far-reaching implications for societal development and environmental stewardship. As AI continues to evolve, its profound impacts across these variables highlight the need for responsible and strategic integration, fostering a sustainable future for both society and the environment.

11.5 RESEARCH METHODOLOGY

In order to thoroughly review and compile a collection of literature on the intersection of artificial intelligence and its role in addressing disaster relief, poverty elevation, and environmental sustainability, this study uses an exploratory research design and a research synthesis approach. While previous studies have primarily examined the benefits of artificial intelligence on social good initiatives, they frequently present a fragmented and linear perspective, concentrating on specific applications rather than offering a thorough and comprehensive understanding of the difficulties and implications involved in using AI to address complex societal issues like environmental sustainability, poverty alleviation, and disaster relief.

This study aims to bridge this gap by adopting an exploratory research design and employing a research synthesis approach. This research aims to provide a more comprehensive view by methodically analyzing and synthesizing the vast body of literature, recognizing the complex nature of AI's contribution to societal good. To fully appreciate the complex issues and possible moral ramifications raised by the application of AI to global issues, a comprehensive perspective is essential. This strategy is anticipated to support informed decision-making and long-lasting solutions by advancing a more nuanced knowledge of the potential and risks involved in using AI for social good.

The research aims to integrate diverse theories and insights from experts in the field, providing a holistic perspective on AI's contributions to social good initiatives in disaster management, poverty reduction, and environmental conservation. To ensure the study's credibility and comprehensiveness, reputable sources, including highly cited documents from databases, such as Google Scholar, are leveraged as the theoretical foundation. This is crucial as the study delves into a relatively novel area that lacks extensive prior research linking AI and its application to social challenges. The selection criteria focus on the past decade, considering the recent emergence of AI as a significant technological paradigm for addressing societal issues.

For source selection, the study included peer-reviewed journal articles, book chapters, conference proceedings, technical reports, and white papers, encompassing a broad scope of applications and outcomes related to AI's impact on disaster relief, poverty alleviation, and environmental sustainability. The studies were limited to those published in English or languages translatable to English, ensuring accessibility to a wider audience.

This methodological approach aimed to provide timely and valuable insights into the contemporary landscape of AI in addressing social challenges, the details of the studies referred are given in Annexure 11.1 (Table 11.1) research studies artificial intelligence for social good, disaster relief, poverty elevation, and environmental sustainability. By grounding the study in reliable and recent sources, it establishes a robust foundation for understanding the intricate relationship between AI and its potential to contribute to social good initiatives. The findings of this study are intended to serve as a valuable resource for policymakers, practitioners, and researchers seeking to harness the full potential of AI in creating innovative solutions for global impact in disaster relief, poverty elevation, and environmental sustainability.

11.6 DISCUSSION

The literature review undertaken indicates that there are both positive impacts of using artificial intelligence and challenges that can arise when artificial intelligence is used. Table 11.1 synthesizes the impact and challenges of the use of artificial intelligence for social good, disaster relief, poverty elevation, and environmental sustainability. Artificial intelligence is making a profound impact across various domains, contributing significantly to social good, disaster management, poverty alleviation, and environmental sustainability. In the realm of AI and social good, projects are actively working toward achieving global goals, building supportive structures, and incorporating diverse perspectives to address societal challenges. AI's positive impact extends to everyday life, influencing healthcare, education, security, crisis management, and promoting social equality. Engaging with communities on a personal level allows for a better understanding of technology, its limits, and potential benefits. Looking ahead, the future impact of AI lies in its practical deployment, addressing global goals outlined by the United Nations Sustainable Development Goals (SDGs).

In disaster management, AI plays a crucial role across various phases, enhancing coordination between human and machine intelligence. The value of AI models in disaster management is evident in providing efficient responses and mitigation strategies. Geospatial technology, including Geographic Information System (GIS) and Remote Sensing (RS), proves vital for understanding emergencies. AI-empowered

TABLE 11.1
Impact and Challenges of Use of AI for Social Good Disaster Relief Poverty Elevation and Environmental Sustainability

Variables	Impact	Challenges
AI and Social Good	AI Projects for Global Goals	Effectiveness and Impact
	Building Supportive Structures	Real-World Impact of AI4SG
	Including Different Perspectives	Ethical Challenges and Potential Overuse
	AI's Positive Impact in Everyday Life	Tactful AI Requires Context Sensitivity
	Engaging with Communities	Limitations in Understanding and Reproducing Success
	Future Impact through Deployment	Ethical Considerations and Biases in AI
	Addressing UN SDGs through AI	Addressing Bias, Transparency, and Fairness
AI in Disaster Management	AI Applications in Disaster Phases	Unacceptable Predictions Beyond Scope
	Coordination of Human and Machine Intelligence	Real-Time Challenges in Emergency Response
	Value of AI Models in Disaster Management	AI's Potential in Disaster Management
	Geospatial Technology and GIS/RS	Success Tied to Data Management and Analytics
	AI-Empowered Lenders and Disaster Mitigation	Focus on Delinquency Outcome and Indicators
	AI Algorithms for Quick Analysis	Geospatial Connection in AI Application
AI for Poverty Elevation	AI for Poverty Prediction	Data Collection Challenges
	Impact on Data Collection	Limited Application of AI in Poverty Prediction
	Role in Agriculture, Education, and Finance	Call for Increased Investment in AI
	Government Investment in AI	Understanding the Impact of AI on SDGs
	AI's Contribution to Poverty Reduction	Opening the Black Box of AI for Social Good
	Influence on Government Decisions	Acknowledgment of AI Limitations
	Educational Preparation for Sustainable Development	
AI for Environmental Sustainability	Foundational Role of AI in Eco-Cities	Limitations of AI in Smarter Eco-Cities
	AI's Broad Impact on SDGs	Specific Challenges for Environmental Sustainability
	Renewable Energy Sector Applications	Future Research Focus
	Consumer Purchase Intentions	Consumer Behavior and Perception
	Intelligent Waste Management	Challenges in Object Recognition
	AI in Farming and Resource Extraction	Governing AI Risks for Sustainability

lenders contribute to disaster mitigation efforts, demonstrating positive outcomes. Additionally, AI algorithms facilitate quick analysis, enabling near real-time results for effective disaster response.

AI for poverty elevation leverages AI for poverty prediction, impacting data collection and revolutionizing sectors like agriculture, education, and finance. Government investment in AI is recommended to achieve Sustainable Development Goals (SDGs) related to innovation, infrastructure, and poverty reduction. AI's contribution to poverty reduction is noteworthy, assisting in mapping poverty concentrations and enhancing the success of poverty reduction initiatives. The influence of AI on government decisions emphasizes the need for educational preparation focused on sustainable development and poverty alleviation.

In the domain of environmental sustainability, AI plays a foundational role in eco-cities, contributing broadly to SDGs and finding applications in the renewable energy sector. AI influences consumer purchase intentions, particularly in autonomous environmental benefits, and holds promise for intelligent waste management. Previous studies (Ovais et al., 2023) have suggested that the hindrance to the necessary behavioral change for preventing environmental degradation lies not in a careless attitude but in a lack of careful consideration. Addressing this challenge, AI holds promise in facilitating behavioral modification to mitigate environmental concerns. AI applications in farming and resource extraction are recognized for their potential to enhance efficiency and productivity. As AI continues to evolve, its impact on these critical variables is shaping a more sustainable and technologically advanced future.

On the other hand, we see the challenges of using artificial intelligence too. The implementation of artificial intelligence is not without its challenges, and these challenges vary across different domains. In the realm of AI and social good, challenges include ensuring the effectiveness and real-world impact of AI projects, addressing ethical concerns, and avoiding potential overuse. Tactful AI deployment requires context sensitivity, and there are limitations in understanding and reproducing successful outcomes. Ethical considerations and biases in AI systems are significant hurdles that need to be addressed, emphasizing the importance of tackling bias, ensuring transparency, and promoting fairness.

In the context of AI in disaster management, challenges arise in handling predictions beyond the algorithm's scope, especially when they are deemed unacceptable. Real-time challenges in emergency response pose difficulties in acquiring actionable and tactical information promptly. Success in disaster management through AI is closely tied to effective data management and analytics. Additionally, focusing on delinquency outcomes and indicators is crucial for assessing the impact of AI on disaster-related scenarios. The geospatial connection in AI applications also presents its own set of challenges.

In AI for poverty elevation, challenges involve data collection, especially in countries facing difficulties in collecting sufficient data for poverty prediction. The limited application of AI in poverty prediction is noted, with the study emphasizing the need for increased investment in AI to unlock its potential benefits. Understanding the impact of AI on Sustainable Development Goals (SDGs) and opening the "black box" of AI for social good are areas that require attention. Acknowledging the limitations of AI is crucial for making informed decisions in poverty alleviation efforts.

For AI in environmental sustainability, challenges include the limitations of AI in smarter eco-cities, specific hurdles related to environmental sustainability, and the need for future research focus. Understanding consumer behavior and perception regarding AI's impact on environmental sustainability is an evolving challenge. Additionally, challenges in object recognition, governing AI risks for sustainability, and addressing biases are critical considerations in ensuring responsible and effective AI implementation in environmental initiatives.

11.7 CONCLUSION

In conclusion, the literature review presents a dual perspective on the use of artificial intelligence, highlighting both its positive impacts and the challenges it poses across various domains. AI is undeniably a transformative force, leaving a profound imprint on social good, disaster management, poverty alleviation, and environmental sustainability. In social good initiatives, AI projects actively contribute to global goals, building structures that support societal challenges, and positively impacting everyday life by influencing sectors like healthcare, education, and security. The engagement with communities and the future practical deployment of AI hold promises for addressing United Nations Sustainable Development Goals (SDGs).

However, alongside these positive impacts, challenges loom in the implementation of AI. In AI and social good, ensuring the effectiveness and real-world impact of projects, addressing ethical concerns, and avoiding potential overuse emerge as challenges. Context sensitivity in AI deployment, limitations in understanding successful outcomes, and the persistence of biases in AI systems underscore the hurdles that must be navigated.

In disaster management, challenges include handling predictions beyond the algorithm's scope and real-time difficulties in emergency response. Success in this domain is intricately linked to effective data management and analytics, emphasizing the need for precision in assessing delinquency outcomes and indicators. Challenges in object recognition, governing AI risks for sustainability, and addressing biases are critical considerations in ensuring responsible and effective AI implementation in environmental initiatives.

Acknowledging the challenges is a crucial step in harnessing the full potential of AI. Moving forward, addressing these challenges is paramount for realizing the benefits of AI while mitigating risks and ensuring ethical and equitable outcomes in its application across diverse sectors.

11.8 IMPLICATIONS OF THE STUDY

The implications of this study are multifaceted and extend to various sectors, policymaking, and the advancement of artificial intelligence for social good. The insights drawn from the literature review on artificial intelligence highlight both its significant positive impacts and the challenges that come with its implementation across various domains. To leverage the benefits of AI effectively, there are several implications that emerge. Organizations and policymakers should strategically deploy AI, especially in areas such as social good, disaster management, poverty alleviation, and environmental sustainability. Alongside deployment, there is a pressing need to prioritize ethical considerations, ensuring fairness, transparency, and accountability in AI systems. Continued investment in research and development is crucial. This involves not only advancing the capabilities of AI but also addressing challenges such as biases, transparency issues, and limitations in understanding and reproducing successful outcomes. Enhancing community engagement and education is vital, particularly in the realm of social good. Bridging the gap between AI researchers and local communities can lead to a better understanding of technology, its limits, and its potential benefits. Educational institutions play a key role in preparing students for a future that incorporates sustainable development and poverty alleviation through AI. Governments play a pivotal role in fostering the responsible use of AI. Support and legislation should be geared toward encouraging innovation, infrastructure development, and poverty reduction. Legislative frameworks should adapt to the evolving landscape of AI to ensure positive societal impacts. In disaster management, precision is crucial. Effective data management and analytics, real-time response capabilities, and a focus on indicators are essential to maximize the benefits of AI in mitigating the impact of disasters. Given the diverse applications of AI, cross-disciplinary collaboration is necessary. Collaboration between technologists, social scientists, policymakers, and environmental experts can lead to more comprehensive and effective solutions. Efforts to mitigate biases in AI systems and promote responsible AI practices should be at the forefront. This involves addressing biases in data, ensuring interpretability and fairness, and governing AI risks, especially in applications related to environmental sustainability. Due to the evolving nature of AI, continuous monitoring and adaptation are crucial. Regular assessments of AI's impact, both positive and negative, should inform adjustments in strategies, regulations, and applications. The assessment of AI's impact on social good, disaster relief, poverty alleviation, and environmental sustainability provides valuable insights for organizations and agencies working in these domains. Understanding the interconnectedness of these societal challenges enables stakeholders to develop more effective and integrated solutions, leveraging AI's capabilities for positive social impact. This can guide the development of AI applications that address specific issues, contributing to sustainable development goals.

By taking these implications into account, stakeholders can navigate the challenges posed by AI while harnessing its transformative potential for the greater good of society. Balancing innovation with responsibility is key to ensuring that AI contributes positively to global goals and sustainable development. Furthermore, the study's exploration of the AI talent landscape through a systematic literature review and bibliometric analysis has implications for the research community. It highlights the current focus areas and engagement of AI researchers, offering a roadmap for key stakeholders in leading research labs to empower researchers further. This insight encourages collaboration and directs research efforts toward addressing the world's most pressing challenges. In conclusion, the implications of this study span societal, policy, and research domains. By providing a nuanced understanding of AI's historical context, current challenges, and potential societal benefits, the research serves as a valuable resource for stakeholders aiming to harness AI's full potential for positive global impact.

ANNEXURE 11.1

S. No	Studies	Journal Name	Authors	Country and Year	AI	Social Good	Disaster Relief	Poverty Elevation	Environmental Sustainability	Key Words Used
1	AIDR: Artificial intelligence for disaster response	Proceedings of the 23rd international conference on world wide web	Imran, M., Castillo, C., Lucas, J., Meier, P., & Vieweg, S.	2014						Stream processing, Crowdsourcing, Classification, Online Machine learning
2	Artificial intelligence and social media to aid disaster response and management	Qatar Foundation Annual Research Conference Proceedings	Imran, M., Alam, F., Ofli, F., & Aupetit, M.	2018						
3	Artificial intelligence for social good		Hager, G. D., Drobnis, A., Fang, F., Ghani, R., Greenwald, A., Lyons, T., … & Tambe, M	2019						AI, Sustainability, Health, Public Safety
4	Artificial intelligence and geospatial analysis in disaster management	The International Archives of the Photogrammetry, Remote Sensing and Spatial Information Sciences,	Ivić, M.	2019						Artificial intelligence, machine learning, geospatial analysis, disaster management, remote sensing
5	Artificial intelligence for social good		Shi, Z. R., Wang, C., & Fang, F.	2020						AI4SG, Environment, Community, Crop Management, Education
6	Harmonizing artificial intelligence for social good	Philosophy & Technology	Berberich, N., Nishida, T., & Suzuki, S.	2020						Ethics of AI · Harmony · Takt · Human-technology interaction · Technological mediation
7	Applications of artificial intelligence for disaster management	Natural Hazards	Sun, W., Bocchini, P., & Davison, B. D.	2020						Disaster resilience, Disaster management, Artificial intelligence

(*Continued*)

(Continued)

S. No	Studies	Journal Name	Authors	Country and Year	AI	Social Good	Disaster Relief	Poverty Elevation	Environmental Sustainability	Key Words Used
8	Artificial Intelligence and Poverty Alleviation: A Review	Mhlanga, D.	Book	2020						
9	A review of explainable AIin the satellite data, deep machine learning, and human poverty domain	Journal of Physics: Conference Series	Isnin, R., Bakar, A. A., & Sani, N. S.	2020						
10	A definition, benchmark and database of AI for social good initiatives	Nature Machine Intelligence	Cowls, J., Tsamados, A., Taddeo, M.	2021						AI, Social Good, AI4SG, Environment
11	How to design AI for social good: seven essential factors.	Ethics, Governance, and Policies in Artificial Intelligence,	Floridi, L., Cowls, J., King, T. C., & Taddeo, M.	2021						AI4SG · Artificial intelligence · Ethics · Social good · Transparency · Privacy · Safety
12	Toward an integrated disaster management approach: how artificial intelligence can boost disaster management.	Sustainability	Abid, S. K., Sulaiman, N., Chan, S. W., Nazir, U., Abid, M., Han, H., … & Vega-Muñoz, A	2021						Disaster management; artificial intelligence; geographic information system
13	Artificial Intelligence (AI) and Poverty Reduction in the Fourth Industrial Revolution(4IR	Journal of Business Research	Ferreira, M. B., Pinto, D. C., Herter, M. M., Soro, J., Vanneschi, L., Castelli, M., & Peres, F.	2021						
14	Utilities of Artificial Intelligence in Poverty Prediction: A Review	Sustainability	Mhlanga, D. (2020).	2021						

15	Artificial intelligence-enabled environmental sustainability of products: Marketing benefits and their variation by consumer, location, and product types	Journal of Cleaner Production,	Frank, B.	2021			Artificial intelligence, Autonomy, Environmental sustainability, Corporate social responsibility, Green purchasing, Robotics
16	Digitalization, circular economy and environmental sustainability: The application of Artificial Intelligence in the efficient self-management of waste.	Sustainability,	Nañez Alonso, S. L., Reier Forradellas, R. F., Pi Morell, O., & Jorge-Vázquez, J.	2021			deep learning; recycling; sustainable self-recycling; convolutional networks; transfer learning; Keras; data augmentation
17	Artificial intelligence, systemic risks, and sustainability.	Technology in Society	Galaz, V., Centeno, M. A., Callahan, P. W., Causevic, A., Patterson, T., Brass, I., … & Levy, K.	2021			Artificial intelligence, Climate change, Sustainability, Systemic risks, Anthropocene, Resilience, Social-ecological systems, Automation, Digitalization
18	Artificial Intelligence in the Industry 4.0, and Its Impact on Poverty, Innovation, Infrastructure Development, and the Sustainable Development Goals: Lessons from Emerging Economies?	Sustainability	Usmanova, A., Aziz, A., Rakhmonov, D., & Osamy, W.	2022			

(Continued)

(Continued)

S. No	Studies	Journal Name	Authors	Country and Year	AI	Social Good	Disaster Relief	Poverty Elevation	Environmental Sustainability	Key Words Used
19	Artificial intelligence for social good: the way forward.		Oliver, N.	2023						AI ü Data authentication ü Healthcare ü Education ü Security ü Crisis management ü Infrastructure administration ü Social equality ü Machine learning ü Personalized treatment ü Adaptive learning ü Threat detection ü Emergency response ü Resource allocation ü Bias mitigation
20	Smart Natural Disaster Relief: Assisting Victims with Artificial Intelligence in Lending.	Information Systems Research.	Liu, Y., Li, X., & Zheng, Z	2023						AI, Natural Disasters, Lending, Delinquency, Credit Scoring, FinTech
21	Deep learning and artificial intelligence in sustainability: a review of SDGs, renewable energy, and environmental health.	Sustainability,	Fan, Z., Yan, Z., & Wen, S.	2023						Artificial intelligence (AI); deep learning (DL); sustainability; renewable energy; environmental health; regulatory oversight
22	Smarter eco-cities and their leading-edge artificial intelligence of things solutions for environmental sustainability: A comprehensive systematic review.	Environmental Science and Ecotechnology	Bibri, S. E., Krogstie, J., Kaboli, A., & Alahi, A.	2024						Smarter eco-cities, Smart eco-cities, Smart cities, Artificial intelligence, Artificial intelligence of things, Machine learning, Environmental sustainability, Climate change

23	Artificial Intelligence (AI) and Poverty Reduction in the Fourth Industrial Revolution.(4IR)	Pre prints	Mhlanga, D.	South Africa (2020)						Artificial intelligence; Fourth Industrial Revolution; Poverty
24	How to Design AI for Social Good: Seven Essential Factors	Science and Engineering Ethics	Floridi, L., Cowls, J., King, T.C. et al.	UK (2020)						AI4SG · Artificial intelligence · Ethics · Social good · Transparency · Privacy · Safety
25	How Artificial Intelligence can help in poverty Reduction in India.	International Journal of Social Science & Economic Research	Rudra Tiwari	India (2021)						Poverty, Agriculture, Education, Corruption, Natural Disaster, Investment Opportunities
26	Environmental Sustainability with Artificial Intelligence	PRA International Journal of Multidisciplinary Research	Manish Yadav, Gurjeet Singh	Multiple counties (2023)						Sustainable Environment, Artificial Intelligence, Environmental sustainability, Environmental sustainability with AI
27	The role of artificial intelligence in achieving the Sustainable Development Goals	Nature Communications	Vinuesa, R., Azizpour, H., Leite, I. et al.	Multiple counties (2020)						Artificial Intelligence, SDG, Climate Action, Poverty.
28	Artificial Intelligence and Poverty Alleviation: A Review	Australasian Conference on Information Systems 2022, Melbourne	Ananya Hadadi Raghavendra, Siddharth Gaurav Majhi, Arindam, Arindam Mukherjee, Pradip Kumar Bala.	Multiple Developed countries (2022)						Artificial intelligence, poverty, sustainable development goals, bibliometric review
29	The Use of Artificial Intelligence in Disaster Management - A Systematic Literature Review	*International Conference on Information and Communication Technologies for Disaster Management*	Nunavath, V., & Goodwin, M	Multiple Developed Countries (2019)						

(*Continued*)

(Continued)

S. No	Studies	Journal Name	Authors	Country and Year	AI	Social Good	Disaster Relief	Poverty Elevation	Environmental Sustainability	Key Words Used
30	AI for social good: unlocking the opportunity for positive impact	Nature Communications	Tomašev, N., Cornebise, J., Hutter, F., Mohamed, S., Picciariello, A., Connelly, B., … Clopath, C. (2020).	Multiple Developed countries (2020)						Machine Learning, Artificial Intelligence, Sustainable Development Goals, NGOs.
31	The Use of Artificial Intelligence for Disasters	Open Journal of applied Sciences	Asma Salman Alruqi, Mehmet Sabih Aksoy	SaudiArabia (2023)						Artificial Intelligence, Disaster Management, GIS
32	A definition, benchmark and database of AI for social good initiatives	Nature Machine Intelligence	Cowls, J., Tsamados, A., Taddeo, M. et al.	UK (2021)						Sustainability Development Goals, AI4SG (Artificial Intelligence for Social Good), Mitigate Environmental Risks, Disaster Management, Climate action.
33	AI for social good: Whose good and who's good? Introduction to the special issue on AI for social Good.	Philosophy & Technology	Josh Cowls	UK (2021)						Artificial Intelligence, Social good, Memory, Intelligence, deep natural language.
34	Artificial intelligence and sustainable development goals nexus via four vantage points	Technology in Society	Osama Nasir, Rana Tallal Javed, Shivam Gupta, Ricardo Vinuesa, Junaid Qadir	2023						Artificial Intelligence, AI-Ethics Sustainable development goals AI for SDG

REFERENCES

Abid, S. K., Sulaiman, N., Chan, S. W., Nazir, U., Abid, M., Han, H., … & Vega-Muñoz, A. (2021). Toward an integrated disaster management approach: how artificial intelligence can boost disaster management. *Sustainability, 13*(22), 12560.

Alruqi, A. S., & Aksoy, M. S. (2023). The Use of Artificial Intelligence for Disasters. *Open Journal of Applied Sciences, 13*(5), 731–738.

Bachmann, N., Tripathi, S., Brunner, M., & Jodlbauer, H. (2022). The contribution of data-driven technologies in achieving the sustainable development goals. *Sustainability, 14*(5), 2497.

Barrat, J. (2023). *Our final invention: Artificial intelligence and the end of the human era.* Hachette UK.

Berberich, N., Nishida, T., & Suzuki, S. (2020). Harmonizing artificial intelligence for social good. *Philosophy & Technology, 33*, 613–638.

Bibri, S. E., Krogstie, J., Kaboli, A., & Alahi, A. (2024). Smarter eco-cities and their leading-edge artificial intelligence of things solutions for environmental sustainability: A comprehensive systematic review. *Environmental Science and Ecotechnology, 19*, 100330.

Cowls, J. (2021). 'AI for Social Good': Whose Good and Who's Good? Introduction to the Special Issue on Artificial Intelligence for Social Good. *Philosophy & Technology, 34*(Suppl 1), 1–5.

Cowls, J., Tsamados, A., Taddeo, M., & Floridi, L. (2021). A definition, benchmark and database of AI for social good initiatives. *Nature Machine Intelligence, 3*(2), 111–115.

de Almeida, P. G. R., dos Santos, C. D., & Farias, J. S. (2021). Artificial intelligence regulation: a framework for governance. *Ethics and Information Technology, 23*(3), 505–525.

Fan, Z., Yan, Z., & Wen, S. (2023). Deep learning and artificial intelligence in sustainability: a review of SDGs, renewable energy, and environmental health. *Sustainability, 15*(18), 13493.

Ferreira, M. B., Pinto, D. C., Herter, M. M., Soro, J., Vanneschi, L., Castelli, M., & Peres, F. (2021). Using artificial intelligence to overcome over-indebtedness and fight poverty. *Journal of Business Research, 131*, 411–425.

Floridi, L., Cowls, J., Beltrametti, M., Chatila, R., Chazerand, P., Dignum, V., … & Vayena, E. (2021). An ethical framework for a good AI society: Opportunities, risks, principles, and recommendations. *Ethics, Governance, and Policies in Artificial Intelligence*, 19–39.

Floridi, L., Cowls, J., King, T. C., & Taddeo, M. (2021). How to design AI for social good: seven essential factors. *Ethics, Governance, and Policies in Artificial Intelligence*, 125–151. https://doi.org/10.1007/978-3-030-81907-1_9

Frank, B. (2021). Artificial intelligence-enabled environmental sustainability of products: Marketing benefits and their variation by consumer, location, and product types. *Journal of Cleaner Production, 285*, 125242.

Galaz, V., Centeno, M. A., Callahan, P. W., Causevic, A., Patterson, T., Brass, I., … & Levy, K. (2021). Artificial intelligence, systemic risks, and sustainability. *Technology in Society, 67*, 101741.

Hager, G. D., Drobnis, A., Fang, F., Ghani, R., Greenwald, A., Lyons, T., … & Tambe, M. (2019). Artificial intelligence for social good. arXiv preprint arXiv:1901.05406.

Imran, M., Alam, F., Ofli, F., & Aupetit, M. (2018, March). Artificial intelligence and social media to aid disaster response and management. In *Qatar Foundation Annual Research Conference Proceedings* (Vol. 2018, No. 3, p. ICTPD1030). Qatar: HBKU Press.

Imran, M., Castillo, C., Lucas, J., Meier, P., & Vieweg, S. (2014, April). AIDR: Artificial intelligence for disaster response. In *Proceedings of the 23rd International Conference on World Wide Web* (pp. 159–162).

Isnin, R., Bakar, A. A., & Sani, N. S. (2020, April). Does Artificial Intelligence Prevail in Poverty Measurement? *Journal of Physics: Conference Series, 1529*(4), 042082.

Ivić, M. (2019). Artificial intelligence and geospatial analysis in disaster management. *The International Archives of the Photogrammetry, Remote Sensing and Spatial Information Sciences*, *42*, 161–166.

Liu, Y., Li, X., & Zheng, Z. (2023). Smart natural disaster relief: Assisting victims with artificial intelligence in lending. *Information Systems Research*. https://pubsonline.informs.org/action/doSearch?AllField=Smart+Natural+Disaster+Relief%3A+Assisting+Victims+with+AI+in+Lending

Mhlanga, D. (2020). *Artificial Intelligence (AI) and poverty reduction in the Fourth Industrial Revolution (4IR)*.

Mhlanga, D. (2021). Artificial intelligence in the industry 4.0, and its impact on poverty, innovation, infrastructure development, and the sustainable development goals: Lessons from emerging economies? *Sustainability*, *13*(11), 5788.

Nañez Alonso, S. L., Reier Forradellas, R. F., Pi Morell, O., & Jorge-Vázquez, J. (2021). Digitalization, circular economy and environmental sustainability: The application of Artificial Intelligence in the efficient self-management of waste. *Sustainability*, *13*(4), 2092.

Nasir, O., Javed, R. T., Gupta, S., Vinuesa, R., & Qadir, J. (2023). Artificial intelligence and sustainable development goals nexus via four vantage points. *Technology in Society*, *72*, 102171.

Nunavath, V., & Goodwin, M. (2019, December). The use of artificial intelligence in disaster management-a systematic literature review. In *2019 International Conference on Information and Communication Technologies for Disaster Management (ICT-DM)* (pp. 1–8). IEEE.

Oliver, N. (2023). *Artificial intelligence for social good: the way forward*. https://research-and-innovation.ec.europa.eu/system/files/2022-07/ec_rtd_srip-2022-report-chapter-11.pdf

Ovais, D. (2023). Students' sustainability consciousness with the three dimensions of sustainability: Does the locus of control play a role?. *Regional Sustainability*, *4*(1), 13–27.

Ovais, D., Simon, R., & Kadeer, N. (2023). Mapping the climate change attitude: careless or care less?. *Environment, Development and Sustainability*, 1–20.

Raghavendra, A. H., Majhi, S. G., Mukherjee, A., & Bala, P. K. (2022). Artificial intelligence and poverty alleviation: A review. In *Australasian Conference on Information Systems. ACIS 2022 Proceedings* (Vol. 78).

Sætra, H. S. (2022). *AI for the sustainable development goals*. CRC Press.

Shi, Z. R., Wang, C., & Fang, F. (2020). Artificial intelligence for social good: A survey. *arXiv preprint arXiv:2001.01818*.

Shneiderman, B. (2020). Bridging the gap between ethics and practice: guidelines for reliable, safe, and trustworthy human-centered AI systems. *ACM Transactions on Interactive Intelligent Systems (TiiS)*, *10*(4), 1–31.

Stahl, B. C. (2021). *Artificial intelligence for a better future: an ecosystem perspective on the ethics of AI and emerging digital technologies* (p. 124). Springer Nature.

Sun, W., Bocchini, P., & Davison, B. D. (2020). Applications of artificial intelligence for disaster management. *Natural Hazards*, *103*(3), 2631–2689.

Taddeo, M., & Floridi, L. (2021). How AI can be a force for good–an ethical framework to harness the potential of AI while keeping humans in control. In *Ethics, governance, and policies in artificial intelligence* (pp. 91–96). Cham: Springer International Publishing. https://www.science.org/journal/science

Tomašev, N., Cornebise, J., Hutter, F., Mohamed, S., Picciariello, A., Connelly, B., ... & Clopath, C. (2020). AI for social good: Unlocking the opportunity for positive impact. *Nature Communications*, *11*(1), 2468.

Usmanova, A., Aziz, A., Rakhmonov, D., & Osamy, W. (2022). Utilities of artificial intelligence in poverty prediction: A review. *Sustainability*, *14*(21), 14238.

Vinuesa, R., Azizpour, H., Leite, I., Balaam, M., Dignum, V., Domisch, S., … & Fuso Nerini, F. (2020). The role of artificial intelligence in achieving the sustainable development goals. *Nature Communications*, *11*(1), 1–10.

Wilkes-Allemann, J., Kopp, M., Van der Velde, R., Bernasconi, A., Karaca, E., Čepić, S., … & Živojinović, I. (2023). Envisioning the future—Creating sustainable, healthy and resilient BioCities. *Urban Forestry & Urban Greening*, *84*, 127935.

Yadav, M., & Singh, G. (2023). Environmental sustainability with artificial intelligence. *EPRA International Journal of Multidisciplinary Research (IJMR)*, *9*(5), 213–217.

12 Digital Innovations for Increasing Financial Inclusion

CBDC, Cryptocurrency, Embedded Finance, Artificial Intelligence, WaaS, Fintech, BigTech, and DeFi

Peterson K. Ozili
Central Bank of Nigeria, Nigeria

12.1 INTRODUCTION

The rising number of people without bank accounts in the world has led policymakers and practitioners to invent innovative ways to increase financial inclusion for over two billion people who are unbanked, globally. Many of these innovations are digital in nature and are mostly the result of private sector intervention while very few digital innovations are led by the public sector. There is high optimism, even among academics, that digital innovations would accelerate financial inclusion by providing an easy, safe, and efficient way to bring formal financial services to marginalized and underserved populations (Chen et al. 2022; Yang et al. 2022; Lee et al. 2023).

The recent COVID-19 pandemic of 2020–2022 accelerated the use of digital innovation to access basic financial services during the COVID-era lockdown and social distancing restrictions in many countries (Eton et al. 2023; Ravishankar et al. 2023). During this time, although many people transitioned to the digital economy, a large segment of the population were both digitally excluded and financially excluded (Gill et al. 2021; Bastick and Mallet-Garcia 2022). This led to concerns about creating a more equitable and inclusive digital and financial system. As a result, several digital innovations have emerged to assist in creating a more inclusive digital and financial system (Malladi et al. 2021; Frost et al. 2021).

This chapter explores some of these digital innovations. They include central bank digital currency, cryptocurrency, embedded finance, artificial intelligence, wallet-as-a-service, Fintech, BigTech, and decentralized finance. By definition, a central bank digital currency is the digital alternative to fiat paper money and is issued by a central

 DOI: 10.1201/9781032644509-12

bank (Infante et al. 2022). Embedded finance is the incorporation of financial products and services into the business platforms of a non-financial services company (Ozili 2022a). Artificial intelligence is the use of computer systems to mimic human intelligence (Haenlei and Kaplan 2019). Wallet-as-a-service refers to access and use of financial services from a customized wallet (Castejon-Molina et al. 2023). Fintech entails access, use, and delivery of financial services using software technology (Puschmann 2017). BigTech refers to large non-financial corporations that offer financial services on their own platforms (Frost et al. 2019; Liu et al. 2022). Decentralized finance (DeFi) describes access to financial services and products from a decentralized blockchain network and without requiring intermediaries (Harvey et al. 2021; Popescu 2022).

These digital innovations are important, and they can be used to support ongoing efforts to increase financial inclusion. However, there is limited knowledge in the literature on the role of these digital innovations in accelerating financial inclusion. The academic literature has not produced significant research on how these digital innovations might accelerate financial inclusion. This might be because these innovations are still emerging or due to lack of a research agenda in this area. Therefore, this chapter attempts to address this gap in the literature, by offering some insight into how these digital innovations might increase financial inclusion. The discussion in the chapter takes a generalist approach, meaning that the discussion is not specific to any country, even though the strategies discussed are applicable to most developing countries.

The contribution of this study to the literature are twofold. First, the discussion in this chapter contributes to the literature that examines the role of digital innovations in improving development outcomes. The study shows that digital innovation can lead to positive development outcomes by accelerating financial inclusion for unbanked and underserved people. Second, the study contributes to the financial inclusion literature that examines the strategies or tools that are useful in accelerating financial inclusion.

The rest of the study has three sections. Section 2 presents the literature review. Section 3 identifies some of the digital innovations for increasing financial inclusion, while Section 4 presents the conclusion of the study.

12.2 LITERATURE REVIEW

Existing studies document some benefits of financial inclusion for societal development. Damodaran (2013) showed that financial inclusion benefits society by decreasing the inequality between the rich and the poor and by making it easier for low income and marginalized people to access the same formal financial services that the rich already have access to. Alnabulsi and Salameh (2021) showed that financial inclusion could be used as a strategy to accelerate economic development. They argued that financial inclusion may create opportunities for job creation, greater financial stability and it can make macroeconomic policies more effective, thereby leading to positive economic development. Valencia et al. (2021) investigated whether financial inclusion can complement societal development in terms of achieving the sustainable development goals. They found a causal association between financial inclusion and achieving the sustainable development goals. Similarly, Sarma and

Pais (2011) observed that higher financial inclusion is correlated with greater human development because greater access and use of finance, over time, lead to better income equality in society. Despite the contribution of financial inclusion to societal development, Mader (2018) warned that financial inclusion strategies could lead to the financializaton of poverty in society, and it may also lead to over indebtedness and may expose vulnerable poor people to risks.

Recently, researchers have examined the role of digital technology in increasing financial inclusion. In the digital financial inclusion literature, for example, Bachas et al. (2018) emphasized that using digital technology could go a long way in delivering financial services to the people who need it the most. The authors further stressed that digital technology could bypass existing barriers that have hindered poor people from accessing essential financial services such as high transaction cost and the onerous paperwork required to open a basic bank account. It was also argued that innovations such as debit cards will not only reduce transaction cost but will also enable faster and quicker access to a bank account. Ferrata (2019) argued that digital instruments may boost financial inclusion for the poor because the weakest and poorest among us can use digital instruments to access financial services to improve their life conditions and it contributes to attaining the sustainable development goals; therefore, digital tools and digital finance can help in leaving no one behind toward greater financial inclusion. Gallego-Losada et al. (2023), in a survey of prior studies, document that the deployment of information and communication technologies (ICT) to achieve financial inclusion has given rise to digital financial inclusion, and digital financial inclusion has created opportunities to expand financial services to unbanked and underserved people which contribute to poverty reduction and social development. Kulkarni and Ghosh (2021) showed that there is much progress in digital financial inclusion in India, and digital financial inclusion is giving women greater social autonomy and financial independence. Despite this, many women in India still face a significant barrier that prevent them from accessing digital financial services and the barrier is the wide gender gap in digital financial inclusion in India. The authors conclude by suggesting that financial inclusion policies should be gender-sensitive to close the gender gap in digital financial inclusion for women in India. Shen et al. (2020) assessed the medium or channel through which financial inclusion is achieved in China and found that financial literacy and usage of digital financial services are the most potent mediums through which financial inclusion is achieved in China. They recommend that policymakers in China should focus on improving consumer's use of digital financial services and increasing their financial literacy level. Dluhopolskyi et al. (2023) assessed the effect of the COVID-19 pandemic on digital financial inclusion. The pandemic led to lockdown which abruptly reduced physical access to financial institutions and reduced access and use of basic financial services. They showed that many people turned to digital innovations during the pandemic in order to access existing financial services, and some of the digital innovations that were used to advance financial inclusion during the pandemic emerged before the pandemic, but the pandemic made them more popular such as artificial intelligence, cryptocurrency, central bank digital currency, the Internet of Things (IoT), and blockchain technology. As a result, the demand for these innovations ballooned during the COVID-19 pandemic and they helped to improve access to financial services.

12.3 SOME RECENT DEVELOPMENT

This section highlights some recent innovative developments that are being used to increase financial inclusion. They are mostly central bank digital currency (CBDC), cryptocurrency, embedded finance, artificial intelligence (AI), wallet-as-a-service (WaaS), Fintech, BigTech, and decentralized finance (DeFi). Some of these digital innovations require internet connectivity to function while others can operate offline as shown in Table 12.1.

12.3.1 Role of Wallet-as-a-Service

Wallet-as-a-service (WaaS) has the potential to increase access to digital financial services and is, therefore, beneficial for digital financial inclusion. WaaS refers to any API-based service or solution that enable unbanked and underserved users to open, manage, and secure a digital account in a wallet. Users of WaaS can use the funds in their wallet to trade and to make an investment digitally to generate income and improve their well-being. WaaS is remarkable because it allows unbanked and underserved adults to seamlessly open a digital account in a wallet, manage their account without needing technical or sophisticated knowledge of blockchain technology, and it enables them to access their wallet from any device at any time and from any location, thereby closing the physical barriers to financial services.

TABLE 12.1
Digital Innovations with Online and Offline Capabilities

Digital Innovations	Can Operate with Internet Connectivity	Can Operate Offline	Remarks
CBDC	✓	✓	CBDC can operate both online and offline
Cryptocurrency	✓		All cryptocurrency users need internet to operate. They need internet to connect with other cryptocurrency users
Embedded finance	✓		Users need internet connection
Artificial intelligence	✓	✓	AI can operate both online and offline because AI systems can interact with the trained data without needing to connect to other AI systems
Wallet as a service	✓		All WaaS platforms need internet to operate
Fintech	✓		All Fintech platforms need internet to operate
BigTech	✓		All BigTech platforms need internet to operate
Decentralized finance	✓		Most blockchain networks and decentralized apps (dApps) rely on the internet to work

Many providers of digital financial services in the private sector are turning to wallet-as-a-service to increase financial inclusion. Deploying WaaS allows providers of digital financial services to incorporate ready-made solutions into a wallet and deliver the wallet to unbanked and underserved adults to meet their unique needs. WaaS offers many advantages for financial inclusion. It presents a reliable solution-in-hand to meet the financial inclusion needs of unbanked and underserved adults as well as for SMEs. It has features that enable the storing of digital money or digital assets. It facilitates seamless cross-border digital payments and transactions at exceptionally low cost. Users can easily obtain the wallet that offer the specific financial services they need, thereby enabling them to access other digital financial services easily.

Despite the potential of WaaS to increase financial inclusion, there are challenges that could hinder the widespread adoption of wallet-as-a-service. One, the presence of a weak or poor digital (blockchain) infrastructure is a challenge. Many countries have a weak or poor digital (blockchain) infrastructure that hinders the widespread use of WaaS. Blockchain infrastructure may be slow and may not offer fast payment to users. Therefore, a weak digital (blockchain) infrastructure may not be able to deploy WaaS effectively. This problem is further compounded by slow internet connectivity. Two, digital literacy and poor knowledge of WaaS is also a challenge. Knowledge of WaaS is still limited among the population. Three, the regulation of WaaS is slow and emerging. This may slowdown the widespread adoption of WaaS. Four, the risk of security breaches is high such as unauthorized access to user wallet that leads to loss of funds in a digital account held in the wallet. This is a significant challenge because users' monies are only as secure as the wallet themselves.

12.3.2 Role of Embedded Finance

Many small businesses are financially excluded. They obtain funding from informal sources. They rely mostly on family, friends, and own funds to raise capital and debt to fund their business. Embedded finance provides an opportunity to change this trend. Embedded finance enables individuals and small business owners to participate in the formal financial system and obtain the funds they need to start and manage their business. Embedded finance entails integrating basic financial services into the product offerings of non-financial firms using APIs (Ozili 2022a). Embedded finance can increase financial inclusion for individuals, small businesses, and MSMEs by giving them the financial tools they need to access funds cheaply, manage their business, and to withstand economic shocks. Many large businesses already offer embedded financial services such as Amazon, Google, Facebook, and Uber. For example, Uber (a non-financial company) integrated a payment and lending service on its app to allow customers make payment without using their own bank app, and the app allows Uber drivers to borrow money (or obtain a loan) from Uber. In the same way, individuals and small businesses can embed basic financial services into their business. This will enable them to provide a convenient and seamless payment experience for their customers at check-out and will also enable them to access cheap and quick funding to start and manage their business. Furthermore, banks and fintech players can reach underserved people and communities by embedding financial services into their mobile devices and in the e-commerce platforms they patronize. This way, banks and fintech players will be able to serve a larger audience and provide financial services to many unbanked adults.

12.3.3 Role of Artificial Intelligence and Robotics

Over the years, banks have been accused of favoring middle-income and high-income customers and abandoning poor and low-income customers in the provision of affordable credit and other financial services. In fact, traditional risk management models and bank regulation encourage banks to avoid doing significant business with poor and low-income customers because of their high-risk profiles, thereby creating inequality in the formal financial system. This inequality is pervasive in the financial services industry because banks do not want to lend to low-end risky customers, and it leads to unequal access to credit and other banking services.

AI and robotics may offer a solution. AI and robotics can help to democratize access to financial services for both unbanked adults, high-end, and low-end customers. AI and robotics can be used to increase financial inclusion for the unbanked by overcoming the challenges of documentation-based tiered know-your-customer (KYC) identification (Kshetri 2021). It allows unbanked adults to use face recognition to meet KYC identification requirements. This will expand access to financial services for unbanked adults in neglected communities. After meeting KYC requirements using face recognition, AI and robotics can also be used to generate alternative non-financial data which can be used to generate credit scores for unbanked adults to enable them access formal credit easily and cheaply (Ozili 2021). Predictive robotics and AI through machine learning can also be used to study customers transaction patterns and generate credit scores based on the observed patterns. Lenders can use the generated credit scores to widen access to formal credit for underserved customers and for unbanked adults in neglected communities (Kshetri 2021). Banks can also use AI and robotics to (i) reduce systemic inequality in the provision of formal financial services to high-end and low-end customers, (ii) change how consumers access their formal accounts and using a more equitable approach, (iii) manage risks, (iv) meet customer needs, and (v) offer affordable credit to unbanked and underserved customers outside of their customer base (Ozili 2021; Kshetri 2021). Despite the potential for AI and robotics to expand financial inclusion, AI and robotics should be deployed carefully to avoid known risks, such as data privacy issues, the replication of societal biases in machine learning models, and over-indebtedness through easy access to digital loans (Kshetri 2021; Ozili 2021; Yasir et al. 2022).

12.3.4 Role of Central Bank Digital Currency

In many countries, central banks are responsible for accelerating financial inclusion and they are constantly in search for effective ways to accelerate financial inclusion. Many central banks are considering using a retail central bank digital currency (CBDC) to increase financial inclusion for unbanked adults. A central bank digital currency is a digital currency or digital money that is issued by a central bank and delivered through an account-based wallet or a token (Auer and Böhme 2020; Ozili 2023b). CBDC may increase financial inclusion in several ways. CBDC can bring many unbanked adults into the formal financial system by first creating digital identities for unbanked adults (Ozili 2023a). CBDC can also increase financial inclusion through ownership of a digital wallet. Individuals without a bank account can own a CBDC digital wallet. They can open a CBDC account to store money digitally, build a credit score, and access other financial instruments or financial services available on the CBDC platform (Allen et al. 2022; Ozili 2023b). Owning a CBDC account does not require burdensome documentation, which makes it easier to enroll compared to the burdensome documentation required by traditional banks (Ozili 2023a). Also, the deployment of offline CBDC will increase financial inclusion for unbanked adults who lack access to internet connectivity either due to lack of internet infrastructure or due to the high cost of internet broadband (Ozili 2023a). To succeed, central banks must determine the best use case of CBDC for financial inclusion and they should ensure that the CBDC is designed with features that increase financial inclusion for unbanked adults.

12.3.5 Role of BigTech

BigTech emerged from the need of non-financial firms to offer financial services. BigTech refers to large non-financial companies that operate platforms that are used by a very large number of people and businesses over the internet (Stulz 2019; Frost et al. 2019; Beck et al. 2022; Ozili 2023c). BigTech can be found mainly in e-commerce (i.e., Amazon and eBay), social media (i.e., Twitter, Facebook, and YouTube), internet search (Google and Yahoo), mobile phone hardware and software (i.e., Apple, IOS, and Android), ride hailing (i.e., Uber and Lyft), and in telecommunications (Frost et al. 2019). BigTech increases financial inclusion by allowing firms to access the financial services and products offered on their platforms (Ozili 2023c). BigTech uses Big Data and machine learning to generate credit scores for clients that want to access the loans offered on the platforms of BigTech firms (Stulz 2019). As a result, they provide convenience and eliminate the need for their clients to visit a physical bank (Frost et al. 2019). BigTech also offer working capital in the form of loans to SMEs to meet their liquidity needs and to help them withstand unexpected financial shock, thereby increasing access to finance for SMEs which is beneficial for financial inclusion (Beck et al. 2022; Ozili 2023c). However, the use of BigTech to advance financial inclusion is challenging because they pose significant financial stability risks, unfair competitive advantage, and data governance risks. In other words, BigTech firms may become too big to fail which would be detrimental for users on their platforms if they fail. BigTech firms also have competitive advantages over smaller firms, and they may kick out small players in that space. There are also data governance issues, such as when collected personal data are not protected or when collected personal data are used to manipulate consumer preferences.

12.3.6 Role of Fintech

Fintech is also another exciting innovation that has huge potential to accelerate financial inclusion. Fintech, also known as financial technology, entails the use of mobile phones, software, and agent networks to enhance access to formal financial services (Sahay et al. 2020). Many fintech providers offer mobile phone-enabled financial services to ensure equal access and use of financial services for the unbanked and underserved segment of the population (Ozili 2018; Makina 2019). Fintech providers have made it easier to onboard more customers and obtain digital loans simply by uploading minimal documentation or credentials into online digital platforms

(Morgan 2022). Fintech is also advancing financial inclusion by reducing operating costs, lowering transaction costs, increasing efficiency in the delivery of financial services, reaching people in remote and rural areas, and serving unbanked adults that are abandoned by traditional banks (Philippon 2019; Ozili 2018). Despite the potential for Fintech to increase financial inclusion, Fintech poses some risks such as cybersecurity risks, weak regulation, over-lending, data privacy, fraud, financial illiteracy, over-indebtedness, and lack of consumer trust (Philippon 2019; Hollanders 2020; Ozili 2023b).

12.3.7 Role of Cryptocurrencies

Cryptocurrencies are digital currencies that serve multiple purposes. Cryptocurrencies are stored and transferred digitally and do not require physical banking infrastructure which makes them useful in expanding financial services in locations and communities where traditional banking services are unavailable or inaccessible (Ozili 2023b). People without internet connection can access cryptocurrencies, thereby removing geographical barriers (Kim et al. 2022). Also, there is no requirement to own a cryptocurrency account and it can be accessed through mobile phones, thereby making it a useful tool for expanding access to financial services for unbanked and underserved adults who cannot be served by traditional banks (Chung et al. 2023). There is also the view that unbanked adults can use cryptocurrency to make an investment, build wealth, and earn income which would enable them to rise above poverty and live a better life (Ozili 2023b). However, the problems with using cryptocurrency to increase financial inclusion are that cryptocurrencies are not primarily designed for the purpose of serving unbanked adults (Carmona 2022). This is because cryptocurrencies are often expensive, volatile, complex, and unsafe which is contrary to what banked adults really need – they need a safe, stable, reliable, and an inexpensive way to access financial services (Kim et al. 2022).

12.3.8 Role of DeFi

Decentralized finance (DeFi) allows people to access financial services and products over a decentralized blockchain network and without requiring intermediaries (Popescu 2022). DeFi aims to democratize access to financial services using peer-to-peer relationships that do not require financial intermediaries. DeFi may increase financial inclusion by (i) giving banked adults control of their money and data which ensures greater privacy and security, (ii) empowering unbanked adults with digital wallets which they can use to access savings, insurance, and other banking products through their smartphones over the internet (Ozili 2022b), (iii) lowering access barriers and expanding access to DeFi platforms using smartphones and other digital devices, (iv) enabling users to borrow funds or lend their assets on a peer-to-peer basis and without undertaking the traditional loan screening checks, (v) ensuring seamless low-cost cross-border transactions, and (vi) enabling developers to build customized financial services that meet the specific needs of unbanked adults (Popescu 2022). Although DeFi can increase financial inclusion, the risks posed by DeFi include the vulnerability of smart contracts, regulatory uncertainty, and interoperability issues, among others (Popescu 2022; Ozili 2022b).

12.4 CONCLUSION

The purpose of the chapter was to highlight the recent digital innovations that are changing the financial inclusion landscape. These developments are simply the emerging digital innovations that are helping to accelerate financial inclusion in many parts of the world. Some of the recent developments that were identified in this study include central bank digital currency, cryptocurrency, embedded finance, artificial intelligence, wallet-as-a-service, Fintech, BigTech, and decentralized finance. It was shown that each of these digital innovations serve a specific purpose, and they contribute to accelerating financial inclusion in unique ways, even though they all pose some risks that can be mitigated with careful and purposeful regulation. These digital innovations will undoubtedly play an important role in advancing financial inclusion in the future. Therefore, the private sector should be encouraged to develop more low-risk digital financial innovations that can accelerate financial inclusion. Policymakers should support the private sector by developing appropriate regulatory frameworks that can accommodate emerging digital financial innovations that have the potential to expand access to financial services. However, such regulation should mitigate risks without stifling innovation. Other developments that could be explored in a future research study include non-fungible tokens, Internet-of-Things (IoT), and the metaverse.

REFERENCES

Allen, Franklin, Xian Gu, and Julapa Jagtiani. "Fintech, cryptocurrencies, and CBDC: Financial structural transformation in China." *Journal of International Money and Finance* 124 (2022): 102625.

Alnabulsi, Zaynab Hassan, and Rafat Salameh Salameh. "Financial inclusion strategy and its impact on economic development." *International Journal of Economics and Finance Studies* 13, no. 2 (2021): 226–252.

Auer, Raphael, and Rainer Böhme. "The technology of retail central bank digital currency." *BIS Quarterly Review*, March (2020): 85–100. https://ssrn.com/abstract=3561198

Bachas, Pierre, Paul Gertler, Sean Higgins, and Enrique Seira. "Digital financial services go a long way: Transaction costs and financial inclusion." In *AEA Papers and Proceedings*, vol. 108, pp. 444–448. 2014 Broadway, Suite 305, Nashville, TN 37203: American Economic Association, 2018.

Bastick, Zach, and Marie Mallet-Garcia. "Double lockdown: The effects of digital exclusion on undocumented immigrants during the COVID-19 pandemic." *New Media & Society* 24, no. 2 (2022): 365–383.

Beck, Thorsten, Leonardo Gambacorta, Yiping Huang, Zhenhua Li, and Han Qiu. "Big techs, QR code payments and financial inclusion." (2022).

Carmona, Tonantzin. "Debunking the narratives about cryptocurrency and financial inclusion." (2022).

Castejon-Molina, Diego, Dimitrios Vasilopoulos, and Pedro Moreno-Sanchez. "CBDC-cash: How to fund and defund CBDC wallets." *Cryptology ePrint Archive* 2023 (2023): 116.

Chen, Yang, Shengping Yang, and Quan Li. "How does the development of digital financial inclusion affect the total factor productivity of listed companies? Evidence from China." *Finance Research Letters* 47 (2022): 102956.

Chung, Sunghun, Keongtae Kim, Chul Ho Lee, and Wonseok Oh. "Interdependence between online peer-to-peer lending and cryptocurrency markets and its effects on financial inclusion." *Production and Operations Management* 32, no. 6 (2023): 1939–1957.

Damodaran, Akhil. "Financial inclusion: Issues and challenges." *AKGEC International Journal of Technology* 4, no. 2 (2013): 54–59.

Dluhopolskyi, Oleksandr, Olena Pakhnenko, Serhiy Lyeonov, Andrii Semenog, Nadiia Artyukhova, Marta Cholewa-Wiktor, and Winczysław Jastrzębski. "Digital financial inclusion: COVID-19 impacts and opportunities." *Sustainability* 15, no. 3 (2023): 2383.

Eton, Marus, Fabian Mwosi, and Mary Ejang. "The effect of COVID-19 on financial inclusion in the Kigezi and Lango subregions in Uganda." *Journal of the International Council for Small Business* 4, no. 2 (2023): 89–102.

Ferrata, Luigi. "Digital financial inclusion–an engine for "leaving no one behind"." *Public Sector Economics* 43, no. 4 (2019): 445–458.

Frost, J., Gambacorta, L., & Shin, H. S. (2021). From financial innovation to inclusion. *Finance & Development*, 14–17.

Frost, Jon, Leonardo Gambacorta, Yi Huang, Hyun Song Shin, and Pablo Zbinden. "BigTech and the changing structure of financial intermediation." *Economic Policy* 34, no. 100 (2019): 761–799.

Gallego-Losada, María-Jesús, Antonio Montero-Navarro, Elisa García-Abajo, and Rocío Gallego-Losada. "Digital financial inclusion. Visualizing the academic literature." *Research in International Business and Finance* 64 (2023): 101862.

Gill, Whitney, Hara Sukhvinder, and Whitney Linda. "A matter of life and death: How the Covid-19 pandemic threw the spotlight on digital financial exclusion in the UK." In *Information security technologies for controlling pandemics*, pp. 65–108. Cham: Springer International Publishing, 2021.

Haenlein, Michael, and Andreas Kaplan. "A brief history of artificial intelligence: On the past, present, and future of artificial intelligence." *California Management Review* 61, no. 4 (2019): 5–14.

Harvey, Campbell R., Ashwin Ramachandran, and Joey Santoro. *DeFi and the Future of Finance*. John Wiley & Sons, 2021.

Hollanders, Marc. "FinTech and financial inclusion: Opportunities and challenges." *Journal of payments strategy & systems* 14, no. 4 (2020): 315–325.

Infante, Sebastian, Kyungmin Kim, André F. Silva, and Robert J. Tetlow. "The macroeconomic implications of CBDC: A review of the literature." (2022).

Kim, Daehan, Maggie Chen, and Doojin Ryu. "Search-Theoretic approach to cryptocurrency adoption and financial inclusion." Available at SSRN 4116714 (2022).

Kshetri, Nir. "The role of artificial intelligence in promoting financial inclusion in developing countries." *Journal of Global Information Technology Management* 24, no. 1 (2021): 1–6.

Kulkarni, Lalitagauri, and Anandita Ghosh. "Gender disparity in the digitalization of financial services: challenges and promises for women's financial inclusion in India." *Gender, Technology and Development* 25, no. 2 (2021): 233–250.

Lee, Chien-Chiang, Runchi Lou, and Fuhao Wang. "Digital financial inclusion and poverty alleviation: Evidence from the sustainable development of China." *Economic Analysis and Policy* 77 (2023): 418–434.

Liu, Lei, Guangli Lu, and Wei Xiong. The big tech lending model. No. w30160. National Bureau of Economic Research, 2022.

Mader, Philip. "Contesting financial inclusion." *Development and Change* 49, no. 2 (2018): 461–483.

Makina, Daniel. "The potential of FinTech in enabling financial inclusion." In *Extending financial inclusion in Africa*, pp. 299–318. Academic Press, 2019.

Malladi, Chandra Mohan, Rupesh K. Soni, and Sanjay Srinivasan. "Digital financial inclusion: Next frontiers—Challenges and opportunities." *CSI Transactions on ICT* 9, no. 2 (2021): 127–134.

Morgan, Peter J. "Fintech and financial inclusion in Southeast Asia and India." *Asian Economic Policy Review* 17, no. 2 (2022): 183–208.

Ozili, Peterson K. "Impact of digital finance on financial inclusion and stability." *Borsa Istanbul Review* 18, no. 4 (2018): 329–340.

Ozili, Peterson K. "Big data and artificial intelligence for financial inclusion: Benefits and issues." *Artificial intelligence fintech, and financial inclusion* Boca Raton, FL: CRC Press, 2021.

Ozili, Peterson K. "Embedded finance: assessing the benefits, use case, challenges and interest over time." *Journal of Internet and Digital Economics* 2, no. 2 (2022a): 108–123.

Ozili, Peterson K. "Decentralized finance research and developments around the world." *Journal of Banking and Financial Technology* 6, no. 2 (2022b): 117–133.

Ozili, Peterson K. "eNaira central bank digital currency (CBDC) for financial inclusion in Nigeria." In *Digital economy, energy and sustainability: Opportunities and challenges*, pp. 41–54. Cham: Springer International Publishing, 2023a.

Ozili, Peterson K. "CBDC, Fintech and cryptocurrency for financial inclusion and financial stability." *Digital Policy, Regulation and Governance* 25, no. 1 (2023b): 40–57.

Ozili, Peterson K. "Determinants of FinTech and BigTech lending: the role of financial inclusion and financial development." *Journal of Economic Analysis* 2, no. 3 (2023c): 66–79.

Philippon, T. (2019). On fintech and financial inclusion. *National Bureau of Economic Research No. 26330*.

Popescu, Andrei Dragos. "Understanding FinTech and Decentralized Finance (DeFi) for financial inclusion." In *FinTech development for financial inclusiveness*, pp. 1–13. IGI Global, 2022.

Puschmann, T. (2017). Fintech. *Business & Information Systems Engineering*, 59, 69–76.

Ravishankar, Pavan, Sudarsan Padmanabhan, and Balaraman Ravindran. "Financial exclusion of internal migrant workers of India during COVID-19: can digital financial inclusion be facilitated by AI?" *Journal of Information Technology Case and Application Research*, 25, no. 2 (2023): 129–158.

Sahay, Ratna, Ulric Eriksson von Allmen, Amina Lahreche, Purva Khera, Sumiko Ogawa, Majid Bazarbash, and Kimberly Beaton. *The promise of fintech: Financial inclusion in the post COVID-19 era*. International Monetary Fund, 2020.

Sarma, Mandira, and Jesim Pais. "Financial inclusion and development." *Journal of International Development* 23, no. 5 (2011): 613–628.

Shen, Yan, C. James Hueng, and Wenxiu Hu. "Using digital technology to improve financial inclusion in China." *Applied Economics Letters* 27, no. 1 (2020): 30–34.

Stulz, René M. "Fintech, bigtech, and the future of banks." *Journal of Applied Corporate Finance* 31, no. 4 (2019): 86–97.

Valencia, Daniel Cardona, Carola Calabuig, Eliana Villa, and Fray Betancur. "Financial inclusion as a complementary strategy to address the SDGs for society." In *Sustainable development goals for society vol. 1: Selected topics of global relevance*, pp. 79–89. Cham: Springer International Publishing, 2021.

Yang, Xiaolan, Yidong Huang, and Mei Gao. "Can digital financial inclusion promote female entrepreneurship? Evidence and mechanisms." *The North American Journal of Economics and Finance* 63 (2022): 101800.

Yasir, Anam, Alia Ahmad, Sagheer Abbas, Mohammad Inairat, Amer Hani Al-Kassem, and Atta Rasool. "How Artificial Intelligence Is Promoting Financial Inclusion? A Study on Barriers of Financial Inclusion." In *2022 International Conference on Business Analytics for Technology and Security (ICBATS)*, pp. 1–6. IEEE, 2022.

Index

Pages in *italics* refer to figures and pages in **bold** refer to tables.

For Product Safety Concerns and Information please contact our EU representative GPSR@taylorandfrancis.com Taylor & Francis Verlag GmbH, Kaufingerstraße 24, 80331 München, Germany

Batch number: 10397790

Printed by Printforce, the Netherlands